# The Market in Mind

# The Market in Mind

How Financialization Is Shaping Neuroscience, Translational Medicine, and Innovation in Biotechnology

Mark Dennis Robinson

The MIT Press
Cambridge, Massachusetts
London, England

This book was set in ITC Stone Serif Std and ITC Stone Sans Std by Toppan Best-set Premedia Limited.

Library of Congress Cataloging-in-Publication Data

Names: Robinson, Mark Dennis, author.
Title: The market in mind : how financialization is shaping neuroscience, translational medicine, and innovation in biotechnology / Mark Dennis Robinson.
Description: Cambridge, MA : The MIT Press, [2019] | Includes bibliographical references and index.
Identifiers: LCCN 2018039817 | ISBN 9780262536875 (pbk. : alk. paper)
Subjects: | MESH: Translational Medical Research—economics | Neurosciences—economics | Biotechnology—economics | Inventions—economics | Technology Transfer | Commerce | United States
Classification: LCC R850 | NLM W 20.55.T7 | DDC 610.72—dc23LC record available at https://lccn.loc.gov/2018039817

# Contents

Preface    vii
Acknowledgments    xiii

1    Introduction: An Ethnographic Analysis of Translational
     Neuroscience    1
2    The Histories of Translational Science and Medicine: Translation as
     a Political Economic Imperative    29
3    Science as Finance: The Financialization of Translational Science
     and Medicine    49
4    The Bench: Universities and Laboratories under Translation    97
5    Bridging the "Valley of Death": Does Translational Research
     Nurture Innovation?    145
6    The Bedside: Patients and Pragmatics and the Promise of
     Health    203
7    Conclusion    229

Notes    241
References    251
Index    285

# Preface

In 2012, the US National Institutes of Health (NIH) quietly announced a new pilot program, *Discovering New Therapeutic Uses for Existing Molecules*. Managed via its National Center for Advancing Translational Sciences, the NIH did something rather curious. In 2013, it announced that it had "matched" researchers at Pfizer with professors at Yale University in the US to work on a potential therapy for Alzheimer's disease. According to the NIH, the aim of the partnership was "to test ideas for new therapeutic uses, with the ultimate goal of identifying promising new treatments for patients."

Although the NIH had always promoted the commercialization of bioscientific research, such deliberate intellectual public-private matchmaking by the NIH was unprecedented. Press releases about the new partnership garnered relatively little scholarly attention. Even fewer noticed Pfizer's near simultaneous closure of its internal neuroscience research arm. Meanwhile, the National Center for Advancing Translational Sciences (NCATS), the progenitor of the Pfizer and Yale alliance, created a flurry of similar partnerships across the US all organized under the goal of spurring something that by 2013 had become a bona fide paradigm: that of *translational science and medicine*.

This book is about the emergence of translational science and medicine in the West, part of a tectonic shift that has quietly reconfigured the landscape of biomedical research in the West. Despite the many definitions accorded to translational science and medicine (TSM), they share a core thread—that biomedical research must be dramatically reorganized to accelerate the transformation of research discovered in laboratories into medical products, including diagnostics, medicines, and technologies. Beyond mere

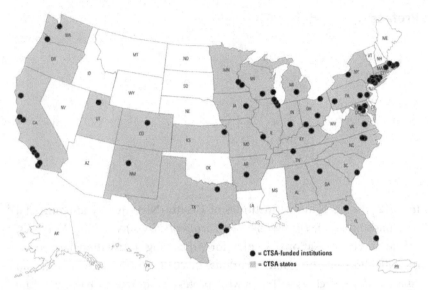

**Figure 0.1**
Clinical and Translational Science Awards (CTSA) site map. By 2017, the number of
TSM-focused centers had grown to nearly 60 institutions. *Source*: http://ncats.nih
.gov/ctsa/about/hubs.

ethos, TSM also refers to a set of realized objects—centers, professorships,
funding schemes, TSM centers, and graduate programs (including nearly
60 new US programs in translational science by 2017—see figure 0.1). For
its proponents, the principal goal of translational medicine was singular: to
bring about a massive transformation in health. According to the NIH and
others, the world's sick are both the *raison d'être* for translational medicine
and its essential test.

My argument is that despite the meanings accorded to it, TSM is not
principally about health. In fact, it is constitutively about something else.
A crucial part of the history of translational medicine lies in the evolution of international biotechnology markets during the global economic
crisis starting in 2004–2005. As universities underwent rapid transformation and increasing privatization and as biopharmaceutical companies
endured a pipeline crisis, exacerbated during the global financial downturn, TSM emerged at precisely the moment of a unique market need. Thus,
we must understand the translational shift in light of highly specific market actions—ones not visible through analyses of translational medicine

that treat it as a mere result of a dearth of innovation in medicine or as a spontaneous and intellectual opportunity born from an overflow of new scientific facts.

The book's primary argument is the following: TSM serves as a means to *de-risk* early-stage, high-risk biotechnology investments for later investment by pharmaceutical partners, investors, and state interests. Through an analysis of corporate externalization programs, I show that TSM, at the social level, enables biopharmaceutical companies to outsource the very riskiest parts of early-stage neuroscience innovation to the nonprofit university—a crucial part of a broader financial strategy. In this paradigm, university research teams become small biotechnology startups and external industry partners such as Pfizer become, in essence, early-stage investment firms. Several questions emerge from this paradigm. What does it mean for research universities to become risk shelters? What happens when universities absorb unknown levels of risk previously belonging to large biopharmaceutical companies? Additionally, beyond the economic issues at work lie implications for the production of scientific knowledge. How do entanglements such as the Yale-Pfizer collaboration contour the resulting science? As translational research directs biomedical research toward specific, predetermined research areas, what is at stake for what becomes knowable and what does not? Which research questions get pursued and which are discarded?

Yet beyond questions about science innovation lies a larger one. TSM has required a panoply of restructurings and unprecedented levels of federal funding. The NIH has spent over $5 billion since 2005 on grants through its translational science awards program alone. For its proponents, we ought to mortgage the pains of translational medicine on glittering promises of a truly transformed global health. In policy edicts and NIH guidance documents, one finds sustained references to the capacity of translational medicine to radically improve patient health—defined as the alleviation of suffering, measurable improvement in wellbeing, and the reduction and elimination of medical diseases. In all of these invocations, one sees a thorough intermingling of the moral and medical, of health and hope. Moreover, as TSM comes to be categorized as medical innovation even before these enterprises have yielded tangible benefits for patients, we are compelled to reflect on the category of health itself as symbolic and moral collateral in a broader global investment strategy. Ultimately, I argue that

TSM reflects the "financialization of health." Thus, rather than considering the Pfizer-Yale collaboration in terms of a more extensive economic and financial encroachment on and in health, one must think of how health can actually function as a means and form of finance.

This book's findings emerge from long-term research, analysis of clinical and scientific research, a study of architectural and material infrastructures designed to spur translational activity, and interviews and laboratory ethnography surrounding brain-based translational medicine. Drawing on literature on pharmaceuticalization (Biehl 2007), biomedicalization (Clarke et al. 2003), and especially the political economy of science (Robinson 2017; Tyfield et al. 2017; Mirowski 2011) and considerations of science in translation (Latour 1983; Callon 1986), this book maps the complex economic and moral nexus that lies underneath translational medicine's visions of innovation and tracks a global transformation in science, technology and medicine. Most importantly, the book's intervention is in bringing finance-specific considerations to recent theorizing about innovation in biomedicine and technoscience—areas that have greatly misunderstood and underestimated the importance of political economy in their analyses. In a context in which research and development, or "innovation," is increasingly synonymous with "mergers and acquisitions," it is imperative to understand how science and technology development is shaped by seemingly unrelated shifts in specific corporate sectors. This shift has vast implications for a whole host of scholars, ranging from those in science and technology studies to those working on research and innovation.

While the book engages with work in the fields of political economy of research and innovation, science and technology studies, and especially, medical anthropology, medical sociology, and perhaps principally, bioethics, there are additional and profound considerations for science and technology policy, health policy and public health, economic anthropology, as well as scholarship focused on biotechnology, research management and innovation studies. A final point is that while this case study asks questions of the larger enterprise of TSM, I am not suggesting that TSM is *nothing more than* the product of a set of political economic configurations. Some may criticize my analysis for being too focused on economic drivers of science and medicine and too (historiographically) inattentive to patients and patient activism. There is indeed a challenge related to capturing the role of patients within the translational paradigm. Part of the marginalization of patients emerges from the fact that under TSM patients are marginalized and, as I argue in the book, reduced to their body parts—which both informs and reflects an "epistemology of parts" that characterizes translational thinking.

To be clear, patients matter. Indeed, patient activism has been integral to new biomedical research areas, including application-oriented ones such as translational medicine. As scholars such as Alondra Nelson and Steven Epstein have noted, local patient activism is a critical space of scientific work and innovation. However, I am suggesting that a clear set of global economic shifts were vital, inextricable drivers in the sudden rise of TSM between 2003 and 2012. This book explores this emergence at various registers—from that of patients, found in the latter part of the book, to those of investors, scientists, and entrepreneurs. By taking an in-depth look at translational innovation (on the ground), one finds a reality that betrays many of the sparkling narratives regarding patients, innovation, and science on which TSM is ushered. At the same time, this analysis also reveals a set of swelling, emerging implications brought about by the global rise of TSM and the power of ethnographic, bioethical, and political economic lenses to uniquely highlight these dimensions.

# Acknowledgments

My initial project sought to understand local transformations of contemporary neuroscience laboratories after the "decade of the brain." My focus quickly transformed when I encountered modes of scientific and clinical reasoning increasingly shaped by a "translational turn." This focus offered a unique opportunity to bring together analytics from anthropology's considerations of medicine, health, and society with the abundant scholarship in science and technology studies, science and technology policy, and the political economy of research and innovation to examine the implications of this "translational turn." Given its explicitly bioethical and technological dimensions, this project would become vital to making sense of the way translational science and medicine affect health and the implications of the increasing entanglement of innovation in technoscience and global finance. Thus, this project quickly became one about emerging models of innovation in biotechnology and their implications.

However, my inquiry quickly implicated larger bioethical questions about emerging and ever-consuming visions of the future of global health. Soon, I understood this project in ethical terms: as part of our ties to collective and social obligations toward patients, communities, and advocates who live in the medical and economic entanglements about which this book is concerned. Accordingly, I humbly dedicate this book to the patients and patient communities from whom I learned, and whose life experiences have informed my theoretical and analytical work.

Drawn out of research at Princeton University, I am most thankful to João Biehl, whose ethnographic sensibility made an indelible lifelong impression and also Carolyn Rouse, whose leadership and insights helped shape the project. I am also appreciative of comments by Lochlann Jain at Stanford, whose work and conversations helped me think about the accidental intersections of culture, social structures, and material worlds. This project also reflects the intellectual imprint of thinking on the political economy of research and innovation. Accordingly, I am indebted to Kean Birch, Rebecca Lave, Samantha Vanderslott, Pierre Delvenne, and David Tyfield. I also thank Steve Woolgar, Bennett Holman, Joseph Gabriel, Lucas Marelli, and Sheila Jasanoff for thoughtful observations about my argument. I am especially grateful for the efforts, thinking, and insights of Phil Mirowski.

I am also appreciative of the Wenner-Gren Foundation for Anthropological Research, Princeton University's Center for Health and Wellbeing, the Center for Neuroscience and Society at the University of Pennsylvania, Princeton's Institute for International and Regional Studies, the Institute for the Recruitment of Teachers, the Wicklander Fellowship for Ethics at DePaul University, the Council on Anthropology and Education, the Gianinno Bassetti and Brocher Foundations, and conference support from Professor Ken Oye at MIT. The support from Katie Helke and the editorial staff at MIT Press has been invaluable.

I also want to express appreciation to the anthropology community at Princeton and especially long-term insights from Noelle Mole Liston. I have benefited from several other intellectual communities, too, including the Rome Conference, including Donald Light, Joel Lexchin, and others. I also benefitted from time spent at the Center for Science, Technology, Medicine & Society at the University of California at Berkeley. While there I learned from Vincanne Adams, Nicholas D'Avella, Diana Wear, Aaron Norton, and Ozzie Zehner. I am indebted to the early insights of David Winickoff and in particular to his caution about the political stakes at work in the federal funding of translational science. I am also indebted to colleagues at the Center for Health Policy and Ethics at Creighton University—Jos Welie, Helen Chapple, John Stone (whose mentorship has been transformational), and Amy Haddad. I am also indebted to the verdant intellectual communities at Harvard, including Rebecca Weintraub Brendel and the Center for Bioethics at Harvard Medical School, and the Petrie-Flom Center for Health Law Policy, Biotechnology, and Bioethics at Harvard Law School.

Lastly, I am most grateful to my shimmering stars, Annie, Dennis, George, Lillian, and Digit.

# 1 Introduction: An Ethnographic Analysis of Translational Neuroscience

## 1.1 The Nascent Utopia of Neurotechnology

"The problem with ethics ... is that it's easy to be a critic and it's extraordinarily difficult to try to create value, to create new ways of helping people. The personal, emotional capital that translational innovation requires is awe inspiring." As Zack Lynch said this, he quickly flashed his smile, though smiling in the way one does when saying something profound. He expressed this to me after I asked the entrepreneur and investor a question about the ethics of translational neuroscience: the field created to quickly transform neuroscience into new innovations and biotechnologies.

His answer cut me off before I had fully finished the question: "Neuroethicists out there ... the problem is that they are so deep [in narrowly defined areas] that they lose some of the real-world contextualization of the actual—how things develop in a broad way." With hints of incredulity, he continued: "Silly bioethical issues ... that's why I'm not an academician. God love them."

Despite his words, I did not construe Zack's statement as a mindless, wholesale dismissal of ethics. In fact, he quickly praised the brilliance of prominent Stanford neuroethicist Hank Greely, who works on the ethics of neuroenhancement. Instead, I interpreted his statement as being about the limits and intellectual investments of classical ethical modes, investments that obviate analyses of the real and unfolding consequences of emerging neurotechnological modernity and the real-world contexts to which they are tethered. He was also—and perhaps more so—problematizing quick and too-easy critiques made about scientists and big pharmaceutical companies.

In one view, Zack's point was about the need for an ethics rooted in context and particularity.

In his early forties, Zack leads the Neurotechnology Industry Organization (NIO) and NeuroInsights, a market analysis firm cofounded with his wife, Casey. (The firm has since closed.) Through NeuroInsights and NIO, Zack touches nearly every aspect of commercial neuroscience from lobbying the federal government to connecting investors and entrepreneurs through NeuroInsights' annual Neurotechnology Investing and Partnering Conferences. Because of his wide reach and deep social networks, he is among the most important social connectors in the Silicon Valley neurotechnology community. Yet, his ever-present self-deprecation and quick wit render him engaging and disarming. In fact, his personal warmth obscures his significant influence. He also has an inimitable physical presence. With coiffed hair, a fresh shave, and an eager smile, his persona reveals optimism and aspiration, a far cry from the cold, East Coast pessimism found among Silicon Valley's Ivy League–educated investors and venture capitalists. His stance and smile reflect a nearly permanent confidence; one is convinced by his statements even before he finishes them.

Zack was the first person I reached out to after relocating to San Francisco in 2009. We had initially planned to meet in February 2010 at his home office, but Casey, who was pregnant, went into labor on the scheduled day. We rescheduled and met on a crisp Friday afternoon at La Boulange, a French café on 24th Street in San Francisco's Noe Valley neighborhood. We sat in front of the café with the sun streaming in. He allowed me to tape our conversation. In meeting Zack, my primary goal was to see what else I could learn about this thing called *translational neuroscience*. I was interested in the logics and contexts that animated this new mode of neuroscience.

What exactly is translational neuroscience? In short, *translational neuroscience* refers to a particular mode of neuroscience research focused on accelerating the development of novel brain technologies such as psychopharmaceuticals, diagnostics, and devices. Born, in part, from a considerable health policy effort spearheaded by the US National Institutes of Health (NIH) in 2005, it was to be more than a plan for the commercialization of university neuroscience research. In a catalyzing article published in the *New England Journal of Medicine*, "Translational and Clinical Science: Time for a New Vision," Elias Zerhouni, then director of the NIH, articulated how translational research would lead to *real transformation* in human health:

> It is the responsibility of those of us involved in today's biomedical research enterprise to translate the remarkable scientific innovations we are witnessing into health gains for the nation. In order to address this imperative, we at the National Institutes of Health (NIH) asked ourselves: What novel approaches can be developed that have the potential to be truly transforming for human health? (Zerhouni 2005, 1622)

Starting in 2006, the NIH led a national transformation of university-based neuroscience centers, creating a network of academic clinical translational science centers focused on optimizing research programs and environments in order to push neuroscience research toward commercialization. Commercialization here means the creation of market-ready products that use this science and includes the entire process involved in bringing these products to market and subsequently producing profits. Thus, while the translational imperative indexes visions about the future of health, translational neuroscience also paves the road for explicit commercial opportunities for universities and encourages a shift regarding the mission of university neuroscience research programs. Importantly, it provides a pathway for investors, as well as pharmaceutical and biotechnology companies, to create partnerships with universities and university laboratories. Thus, my principal question regarding translational neuroscience was about its impact on the commercialization of scientific research in university settings. However, to explore this question, I needed to inhabit the worlds of the corporate life sciences and of biotechnology investors. I had hoped that Zack might help me with this, and indeed he did.

I was introduced over email to Zack by Joe Powers, executive director of the University of Pennsylvania's Center for Neuroscience and Society, where I had participated in a neuroscience training program in the summer of 2009. Joe explained that Zack was someone I had to meet given my interests in translational neuroscience (hereafter TN). I had already heard of Zack well before Joe's suggestion. Zack's name emerged in conversations with other academics and during early conversations with pharmaceutical executives. I had even seen him on television discussing his coauthored book, *The Neuro Revolution: How Brain Science Is Changing Our World*, which explores how emerging technologies in neuroscience may radically transform the world.

This book introduces the world of neurofuturism, a field of inquiry and imagination focused on expansive opportunities and transformations that may occur via an increased understanding of the brain. To be clear, the book is also a work of technological futurism, focused on all of the social areas in which future neurotechnologies should revolutionize human life. For example, it discusses how new brain technologies, including software, will be used to hone financial traders' decision making or how neurotechnologies will be used by dating services and in judicial contexts (in the latter case to examine a subject's risk of criminality and as a means of lie detection). Organized around the notion that the "Neuro Revolution" will comprise a significant economic and historic revolution, Zack places it alongside the agricultural, industrial, and information revolutions—all of which produced dramatic social, cultural, and economic effects. For some reviewers, the book was too hyperbolic, nothing more than unstructured futuristic prognostication. Yet one can also read it as a meditation, a consecration of a moral future slowly unfolding: one where we can improve human decision making, where we can eradicate brain diseases, and where people, now thoroughly understood and demythologized via brain science, can be romantically matched to others who are neurologically, psychologically, and emotionally optimal mates.

Zack begins the book by recounting a telling experience in which he was, in a word, *saved* by biotechnology. While on an international vacation, he had a skiing accident that produced a painful and debilitating back injury. After he'd spent years searching for cures, a neurosurgeon ordered him to undergo a full-body MRI scan in 1996. Zack describes the experience of being fully immersed within the scanner: its cacophony of sounds, the visual claustrophobia. In his narrative, it was the magic of the MRI scanner that enabled the doctor to find the diagnosis that led to his recovery. In my reading, this moment constituted a kind of biotechnological conversion narrative: a thoroughly technological soteriology. There, as he emerged from the scanner, Zack (see figure 1.1) received what he called "an invaluable inkling of the future" (Lynch and Laursen 2010, 2).

While Zack is many things—investor, principal of his firm, entrepreneur—he often discusses his background as an economic geographer studying how and why industries emerge, how they develop, and the reasons they do. He explained precisely what demarcates him from others: speculators, neuroscientists, investors, and investment advisors working in the

**Figure 1.1**
Zack Lynch in his home office, in the basement of his house in San Francisco's Noe Valley neighborhood. He and his wife, Casey Lynch, are the owners and directors of NeuroInsights. Image, *San Francisco Chronicle*. Copyright © 2014 SF Gate.

neurotechnology space. He considers himself a *tracker*. He also articulates the importance of understanding how various stakeholders in the translational system may be so domain-focused that they are not able to clearly see how they are part of broader sociotechnical and historical configurations:

> Specialists are very powerful at being able to tell you what's going on [in their own domain] but they can't tell you about converging technologies and so being a tracker, it's very difficult to get people to—other than venture capitalists and even they have an issue with it—to sit down and talk about drugs, devices, and diagnostics [at large] and you're either a biotech person in your silo or you're a medical device person or you're diagnostics, personalized medicine ... but to me [these larger trends are] the essence ... and you have to take into consideration converging technologies. ... You've got to really embed yourself in as much random stuff as possible ... so you're not missing texture. ... I'm like a massive data collector. (Interview 7, February 12, 2010)

Zack's use of the term *tracker* coalesces with the positionality of the ethnographer, simultaneously on the inside and the outside. And in many cases, this describes the radical heterogeneity of his daily routine. During the annual Neurotechnology Investing and Partnering Conference, Zack adroitly introduces investors to startups, deploying enough neuroscience knowledge to know why one particular startup might be of interest to a given investor. On another day he is in Washington, D.C., lobbying Congress or writing op-ed pieces. In each case, he strategically uses his on-the-ground knowledge to structure arguments about trends, challenges, and larger logics at work within commercial neuroscience or to convince, for example, investors to take an interest in a new crop of neurotechnology startups focusing on devices.

Thus, tracking refers to the vast diversity of Zack's experiences and data sources as well as his use of these experiences in his work. He said to me that "I don't consider myself to have a [high] IQ, but I have this ability to deflect and immerse." So while he runs in many related and interconnected circles—biotechnology startups, private equity executives, pharmaceutical company leaders, Silicon Valley entrepreneurs, neurofuturists and neurogamers, popular neuroscience writers, Bay Area neuroscientists, and "normal" professors—his broad interests do not signal dilettantism. His positionality is firmly rooted in allegiances to economic markets and genuine faith in neuroscience itself. Thus, his commitments and location within the world of neurotechnology mean that his tracking enables him to be an effective and diversified broker of the brain. Given Zack's location in various microworlds, he was a particularly useful guide for me as I sought to develop what for anthropology is an emic understanding (i.e., adopting the meaning perspectives of insiders and using internal logics) of large-scale shifts at work within commercial neuroscience and neurotechnology investments. His perspectives were so contoured by his faith in neurotechnology that his insights and work often offered me useful directions in terms of getting at emic meanings.

Zack and Casey founded NeuroInsights in 2003. NeuroInsights is an organization uniquely positioned in the world of neuroscience, neurotechnology, and life science investment communities; while it provides several services, it is mostly a market information, tracking, and analysis firm. Each year NeuroInsights produces the highly influential *Neurotechnology Industry Report*. The 2015–2016 version of this report, for example, costs $5,700 to purchase and is 650 pages long. According to the product description, "It's the only publication to provide a comprehensive pipeline and market analysis to help investors, companies, and entrepreneurs easily identify opportunities, understand the competitive landscape, determine risks, and understand the dynamics of rapidly changing CNS [central nervous system] markets."[1] The report links networks of entrepreneurs, investors, and biopharmaceutical, device, and biosoftware companies otherwise difficult to access. NeuroInsights also works at the interface of neuroscience and translation in that the firm provides strategic business and advisory services to small neurotechnology companies looking for investments, while also providing advisory services and investment leads for those seeking to invest in neurotechnology companies.

Zack also founded the Neurotechnology Industry Organization (NIO), a trade organization composed of neuroscience-related companies, organizations, university research centers, as well as pharmaceutical and medical device companies. This organization focuses on lobbying efforts around federal research funding as well as advocacy for a host of industry interests in everything ranging from taxation of investment gains to regulatory issues of the US Food and Drug administration (FDA), which must approve neurotechnologies before one can bring them to market. NIO and NeuroInsights produce annual events nationally and internationally designed to create opportunities for partnerships and networking between scientists and investors, and between pharmaceutical executives and biotechnology startups, among others. These Neurotechnology Investment and Partnering Conference constitute crucial spaces where TN "happens."

Thus, Zack is uniquely positioned to lay out the landscape of neurotechnology—touching spaces ranging from laboratory discovery all the way through commercialization. To an outsider, he may be a key figure in neurotechnology. Yet this fails to capture the personal importance of the brain for Zack. When talking about emerging technologies, he speaks quickly and excitedly:

> So let's go into cochlear implants, brain stimulators for Parkinson's, and there's a whole frontier of emerging neurodevices to treat a whole series of things and not only psychiatric [diseases]. One company developed a portable TMS [Transcranial Magnetic Stimulation machine] for migraines that goes beyond magnetism and that goes [in]to light therapies that are noninvasive. There are also all the surgical procedures and so it's where it's all going. (Interview 7, February 12, 2010)

Like the investors with whom I spoke, Zack invokes the future in order to contextualize (and often deproblematize) the present. He says, "The crudeness of 2005–2020 when people look back from 2040 will be like the invention of propeller planes ... it's like, 'can you imagine that we couldn't fly to Hawaii?!'"

I asked Zack about the concept that he'd created called *neurosociety*. After a moment, he provided an explanation:

> Humanity has gone through an agricultural society. Currently we're in an information society where information technologies drive fundamental political economic and social change, and so if you look at history you realize that new societies emerge when new technologies develop and coevolve with sociocultural norms; you get these technoeconomic waves and they give rise to new forms of human society that are different from predecessor societies. My background is as an economic geographer—[studying] how and why industries emerge and how they develop and why they do and the reasons why they do. I coined the term *neurosociety* and what this represents is a society where neurotechnologies begin to transform political and economic relations, sociocultural relations, business methods, personal patterns of interconnectivity, ways of being, ways of seeing, ways of existing, and norms of our lives—[and] ... in many ways [are themselves] radically influenced by this society. That's what the neurosociety means. ...
>
> I mean you have these grand revolutions—agricultural, industrial revolution, then information revolution and then the neurorevolution, within those, you can pick out and tease out even more finely grained patterns of technological development that are tied to the capitalist mode of production ... and so Nikolai Kondratieff, who's a Russian economist—came up with classic analyses of these technoeconomic waves. [Economist] Joseph Schumpeter was a big fan ... he [Schumpeter] came up with the term *creative destruction*. ... and so this idea of these technoeconomic waves, they emerge and then go through a process where low-cost products engender entirely new ways of doing business, they impact industries that already exist, they require new industries which create new modes of work and which create new modes of living. (Interview 7, February 12, 2010)

For Zack, there is a crucial relationship between emerging innovations in neuroscience and economic transformations "tied to the capitalist mode of production." His invocation of "technoeconomic waves" may get at the inextricability of "the social" and "the technical" that early economic historians such as Schumpeter and Karl Polanyi prefigured in their theorizing about massive social transformations emerging out of technologically enabled industrial capitalism (Polanyi 2001). Concepts such as that of the sociotechnical ensemble, which places analytical weight on the *inextricability* of the social and the technical in explaining technosocial change, exemplify this convergence (Bijker 1997). The kind of sociotechnical historiography that Zack uses to explain how "new societies emerge when new technologies develop and coevolve with sociocultural norms" makes his notion of neurosociety legible—in some ways—within recent theorizing regarding the coproductions or mutual constitutions between evolving life science epistemologies and capitalist-economic exchange (Rajan and Leonelli 2013). For Zack, this "technoeconomic wave" that neurotechnology enables isn't merely an exercise in elucidating the vicissitudes of history: it's about a nascent utopia that the commercialization of neuroscience will unveil.

## 1.2 Translational Neuroscience as a Moral Imperative

It was during our conversation in early 2010 that I began to wonder whether TN might not be a useful sphere in which to map a broader set of economic shifts. Little did I know that the domain of translation is tethered to a variety of moral and political concerns. For Zack, an ardent technological evangelist, we should be concerned with what will emerge from the inevitable inequalities increased technological efficiency is likely to create. As with many futurists, Zack's fantasies exist alongside concerns about plebeian publics and global underclasses left behind as a result of technological, genetic, and robotic innovation. As with many libertarian moralities, the problems of inequality are thought to be in the domain of perception. When I ask Zack about the social and political futures that neuroscience will author, his statements become dramatic and prophetic. For him, our impending neurotechnological future will be grand, utopian, and moral. I lean in. He continues:

> I'm gravely concerned about humanity's future. I have been ever since I was a young child. My mother took me on a trip to India when I was like 13, and we meditated for six weeks and on my way back we stopped at Dubai airport ... and I saw this radical difference in the disparity of wealth and [came to the conclusion that] when everyone else figures out what everyone else has there will be a cultural war. The most powerful way that we'll be able to influence humanity on a broad scale ... are these new neurotechnologies—whether they are used for warfare or sophisticated neuromarketing and neuropersuasion or rapid education or training in empathy, to me, this is where humanity must go. (Interview 7, February 12, 2010)

Zack's contention, although dramatic, helps one think beyond a view of TN as merely a highly commercialized mode of neuroscience research or as a simple story of science and commercialization. His sentiments compelled me to think about TN as a means of moral envisioning. A masterful storyteller, he had teased out the moral, emotional imaginaries through which significant investments in biomedical research are made: untold scientific progress, groundbreaking technologies, and a universe of medical cures. And yet, at the same time, this sentimentality masked the on-the-ground reality of TN. Behind its glittering promises lay a complex backstory. According to Zack, while innovations in the brain sciences certainly offer an opportunity to enact more equitable global futures, it's the alleviation of disease burdens where society will experience the more immediate impacts of TN.

In our discussion, his tone regarding TN retains a sense of moral imperative. "There are two billion people suffering from brain-related disorders!" He quickly recites all of the diseases and conditions considered brain-related disorders—everything ranging from Alzheimer's to depression and epilepsy. He explains how the sheer market size attached to CNS[2] causes investors to articulate the brain sciences as a space of both market need and incredible difficulty, especially compared with other kinds of diseases:

> So these are *huge* [his emphasis], massive healthcare market opportunities that are profit driven, and they [investors and companies] are looking for new market opportunities, and this is an especially complex area in which to do science and R&D. It's the last frontier primarily driven by the fact that there is a brain-blood barrier scientifically and it's a complex organ. (Interview 7, February 12, 2010)

He also tells me about the trends in diagnosis: "Psychiatry is a much larger market … depression continues to grow versus Alzheimer's (neurology), [which] is smaller," Zack says. However, he provides a glimpse into the kinds of clinical challenges and underlying scientific complexity that plague CNS, describing the ways brain disorders diverge from models used for other kinds of diseases. As an analyst of neurotechnology markets, Zack explains the risks at work for investments in CNS. It was the first time in our conversation that he used the word *failure*:

> You can't tease out causation [in CNS]. In [prospective therapies for the treatment of] cancer you can see if it's in remission in six weeks … and so you see super high failure rates [in CNS]. The data that I use in my meetings on Capitol Hill is that to get to the average clinical trial for CNS drugs costs $1.6 billion versus $800 million for an average pharmaceutical trial. To send it through clinical trials from

figuring out a compound, ... [what about] the lost profit of those programs that didn't make it?! [CNS] also takes longer. [A] pain med can be seen if it's effective within 30 minutes to know if it's working, whereas for Alzheimer's, it may take 3–5 years, and so tracking people for that long costs hundreds of millions of dollars ... and so it costs more for many areas in CNS [as compared to other sectors]. Another issue is that because we're dealing with the brain we have side effects that influence our behaviors, such as suicidality, and safety agencies have woken up too ... and it's another cost for Pharma development. (Interview 7, February 12, 2010)

Zack's figures and industry numbers come quickly. His congressional lobbying and facility with investors mean that he articulates market sizes, corporate histories, and profit-and-loss margins with both clarity and acuity. All of the challenges he articulated constitute what is commonly referred to in investor circles as the "valley of death" (see figure 1.2). Translational science and medicine (of which TN is part) are created to remedy the "valley of death." As Zack describes it, TN is an attempt to "get a compound out of an academic lab, acquire initial funding, put together a small business team

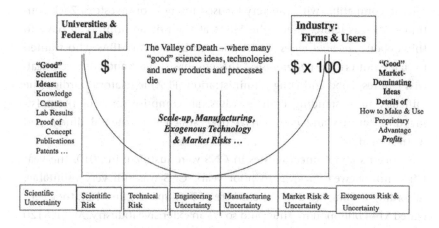

**Figure 1.2**
This graphic helps to explain the various elevations of risk and risk types, as well as the changes in funding availability that co-occur alongside research translation and commercialization. Translational research programs aim to bring more of the "good ideas" toward commercialization. *Source*: "Factors That Foster Industry-University Cooperation: Implications for I/UCRCs." Paper presented by M. Jelinek at the National Science Foundation Industry-University Cooperative Research Center Evaluators' Meeting, Arlington, VA, June 2006. © Mariann Jelinek, PhD, Professor of Strategy Emerita, College of William and Mary, Williamsburg, VA 23185.

because you are a professor focused on translating neuroscience, getting that object out of the marketplace of *ideas* (emphasis mine) into a company. That's the valley of death." As evidenced in his elucidation of the path between research and application, TN is thus envisioned here as a means of materialization, or perhaps a set of materializations.

During this interview, we find an interesting paradox: underneath the triumphalist discourses of TN, the problems at work in CNS co-occur alongside enormous potential profitability and massive risk. The separation between the underlying science of CNS and its (perhaps paradoxical) financial profitability compelled me to think in economic terms about CNS. Zack took the conversation in this direction. He explained how his interests in large-scale socioeconomic shifts spanned his academic and professional interests: "I was doing a PhD in economic geography, and I was trying to understand spatial dynamics. I had to do large-scale economic forecasting and then I was hired by a private firm to do large-scale economic geography." He then mentions that in leaving academia to work for the private sector, he found that his own optimistic neurofuturism was often incompatible with the very focused interests of investors. Zack continued, "I was telling them [the MBAs at the private firm], 'you have to think about *potential forces*' and I got pushback from MBAs who wanted to figure out critical *uncertainties*" (his emphasis). Vast and costly clinical-trial failures, Food and Drug Administration (FDA) regulatory hurdles, the difficulty understanding brain diseases, all comprise some of the existing uncertainties and commercial risks attached to mental illness and its treatment.

Concerns about uncertainties in CNS were justified: in 2010, the year I first interviewed Zack, as in prior years, CNS was still very profitable. According to him, "In 2008, there were 650 companies [in CNS]; they generated $144 billion in revenue, and so it's an extensive industry. About $120 billion was drugs; $20 billion included diagnostics, brain imaging systems, in vitro diagnostics, tests; and $4.5 billion was neurodevices" (Interview 7, February 12, 2010). By 2015, the revenues grew to $172 billion according to the Neurotechnology Industry Organization.

While Zack was outlining matters of concern for neurotechnology investors and market analysts (whom he called "MBAs"), he also underscored the importance of dealing with and managing critical uncertainty (in other words, risk) for companies that worked in CNS. Thus, alongside his optimism, an enormous tension existed in commercial neuroscience: the marshaling of utopian discourses that obscure the riskiness of TN. What I later learned is that there is a complicated and relatively hidden history that helps to explain the sudden explosion and global expansion of translational science and medicine in the 2000s and that continues to shape these areas into the present—a history tied to pharmaceutical divestment, diminishing appetite for shareholder risk, and the emergence of increasingly entrepreneurial universities.

## 1.3   Spaces of Investigation

My research began in a local California TN community in late 2009, almost 10 years after the "decade of the brain." In the 1990s it was predicted that the growing portfolio of discoveries in the brain sciences would allow for unprecedented new knowledge and capacities in the realms of neuroscience, medicine, and society. The gradual unfolding of this promise was to occur in the form of new powers to read the brain or to use fMRI technology to instantly understand psychiatric function and dysfunction.

Yet 2009 was also a moment of economic uncertainty because of the global economic downturn. Thus, it was an exciting time in which to talk with those in the neurotechnology community. I had no idea just how important the issues of economic uncertainty would be for my understanding of the dynamics that undergirded TN. However, my conversations with investors and pharmaceutical executives and the events I attended that focused on investments in the life sciences increasingly addressed risk, financial fallouts, and the lack of any real innovations from CNS—a sector that received significant financial investment dollars.

I spent a total of 24 months in Northern California, where I was a visiting scholar at the Center for Science, Technology, Medicine & Society at UC Berkeley. Following anthropologists who study science outside of the laboratory (Hess 2001; Martin 1991), I explored specific sites of TN: university-based research centers/laboratories, as well as external industry conferences and neurotechnology/neuroscience industry spaces. I also conducted interviews with investors, scientists, clinicians, entrepreneurs, and university administrators, and I spent time in a neuroscience laboratory at a university in Northern California and studied the design of the laboratories, new science campuses, and translation-focused software.

While this project includes several sites of investigation, my research focuses on discourses that transcend particular sites, taking seriously the work of talk around TN in terms of what it accomplishes and what it obscures. Ethnographic perspectives help dislodge some of the more deterministic assumptions about the implications of TN—assumptions that may compel one to think of TN as being primarily about neuroscience and translation or that center ethical issues on access to new neurotechnologies or simple corporate intrusions into science.

I focus on Northern California because of its importance as a premier region in the United States for neurotechnology investment and activity. Eventually, via my relationship with Zack, I was able to inhabit a tight community of companies and biotechnology startups and investors. I also spent time in other spaces outside of Northern California, including time at the Cleveland Clinic in Ohio and at Neurotechnology-focused Conferences. Though the conversations recounted here are verbatim, many of the names associated with them are pseudonyms.

## 1.4   Insights

I spoke with people in various positions—patients, clinicians, investors, neuroscientists, engineers, translational center staff, entrepreneurs—and in learning from my time in a multitude of places and spaces of "translation" as well analyses of media, I came to a key conclusion: TN, and by extension, translational medicine, is not principally about health. This is not to say that TN does not create gains in health. Rather, I demonstrate that TN is better understood as a system of systems enabling the working out of particular problems produced in the corporatization of healthcare. Thus, I argue that in the fallout from the global financial crisis, the emergence of TN intersects with and becomes a solution for the financial costs and risks associated with early-stage neuroscience innovation, risks that actually emerge because of an overleveraging of problematic models in neuropsychiatry. Because of TN's particular financial functionality, I ask the question: Can translational science and medicine increasingly be considered a form of finance?

Given TN's financial operability, I show that university-based TN actually functions as a mechanism at the social level to de-risk high-risk investments for further development by pharmaceutical partners. Given the way that TN enables a transfer of early-stage neuroscience research to universities and Pharma models that seek to externalize innovation sources, I argue that university TN laboratories are becoming research arms of biopharmaceutical companies. In all of these elements, I see the work of what I call the financialization of health taking form. Additionally, as I learned from my time at the Neurotechnology Investing and Partnering Conference, TN is characterized by a unique set of micropractices focused on creating solutions for the problems of biological material.

By facilitating the meeting of a patient's biological material with disease constructions (such as TN's focus on developing diagnostic tools which seek to locate brain disorders), TN seeks to intervene in a science where there are huge knowledge gaps, diseases that are not always well understood, and biotechnologies that don't always work. Thus, at the social level, TN could be thought of as operating on a material-semiotic level, helping produce order in the throes of biological, social, and biotechnological chaos. Building on these arguments, I suggest that TN relies on the creation of epistemological systems that entail the compelling of individual knowledge-actions such as those of the scientist, achieved through the design of novel environments such as the new, innovation-focused science laboratory and software programs designed to compel certain actions around knowledge and information. Lastly, I outline ethical questions and implications associated with translational science and medicine writ large, including questions about the role of the university as a risk shelter, and examine the consequences of "offshoring" the risks of drug development onto the public via the use of federally funded research and partnerships with nonprofit universities

Thus, I offer an alternative to (though not necessarily a refutation of) scholarly treatments of translational science that see it as the coproduction of capitalism and the life sciences or the product of an inevitable evolution of the life sciences. Though elegant, these theorizations obscure or neglect the story of biomedical models falling apart, the crisis in R&D during and after the global economic crisis, the very particular transformations in the biopharmaceutical sector, and especially the impacts of specific transformations in global finance—all of which are necessary (Farquhar and Kelly 2013) to thoroughly map and explain the translational shift. By tracing these specific contextual elements in the rise of TN, I will suggest that a very specific set of documentable histories help explain crucial elements in the rise of TN and translational medicine in general. This history is important. By mapping the crises of commercial neuroscience R&D and drug development, one finds the now bulging outgrowth from biomedicine's many scientific reductions, a fact uncovered by medical anthropologists, sociologists, historians, and others who have mapped how biomedical models reduce important complexities associated with illnesses into "languages" that make them scientifically, technologically, and economically digestible. These "reductions" get embedded or engineered into neurotechnological objects themselves. Shareholder risk during the financial crisis meant a mass exodus from neuroscience R&D and a search for a solution to problems of R&D risk, cost, and shareholder value. That a field such as TN emerged precisely in the aftermath of these failures must be understood as a validation of the numerous critiques of biomedicine in science studies, the critical social sciences, as well as research and innovation scholarship.

Lastly, there may be a question about whether my conclusions about TN are generalizable across other translational fields. TN is one among many translational fields, such as translational oncology, which seeks to turn research into cancer therapies and diagnostics, and translational genomics, which seeks to inject explicit application-oriented aims into genomic research. Like these other fields, TN is part of the larger infrastructure that comprises translational science and medicine. Yet there are many aspects of TN that are specific, given the particularity and challenges of neuroscience research and neuropsychiatric models. Thus, the in-depth analyses in this book are as much a historiography of contemporary neuroscience as they are a picture of science in transformation under financial imperatives. At the same time, these latter imperatives—which are a significant focus of the book—animate the entirety of translational science and medicine as well as contemporary science models writ large. As this project traces both of these poles—the macro and the micro—the book also goes back and forth between this general analytical mode and closer, in-depth considerations.

## 1.5  Knowledge in Translation: Laboratories and the World

The ethical importance of this research for contemporary studies of biomedicine and health lies in its examination of the kinds of claims that can be made in the name of global health as well as the means through which institutional, political, and market processes come to animate notions of innovation and progress in science and medicine. Thus, an exploration of the translational shift must also delve into the very notion of translation—the idea of pulling facts from laboratories into the external world.

Before science and technology studies turned to the laboratory as an object of intellectual interest, scholars tracked the variety of ways knowledge production is embedded in and dependent on "the social." Anthropological research focusing on local, indigenous rationalities (Evans-Pritchard and Gillies 1976) and on the nature and contexts of knowledge, especially in cognitive anthropology (Bateson 2000; D'Andrade 2001; Ruesch and Bateson 1951) and sociology (Zerubavel 1999), provides a rich history of theorizing about the many cultural particularities (Shweder 1991) and inextricable interplays (Bloch 2012) between culture, contexts, and thinking. Mary Douglas, critiquing Émile Durkheim and Marcel Mauss for their differentiation between primitive and modern scientific knowledge-production practices, made the following observation:

> Even in the laboratory, the researcher has options open to him. There are options for following this line of inquiry rather than that, of referring to these other works or omitting them. ... The categories of valuable and useful areas of work are identified, ranked and bounded, elements assigned to classes and sub-classes, rules made to hold the framework of knowledge steady. The alleged gap between what we know about the construction of everyday knowledge and the construction of scientific knowledge is not so big as is supposed. (Douglas 1973, 12)

The sense that all knowledge is equally constituted via thoroughly social means made knowledge of every variety conducive to anthropological inquiry. The ethnographic turn in science and technology studies that focused on everyday laboratory practices and that led to an understanding of how scientific knowledge is collaboratively created follows from a discrete but rich tradition of first-generation laboratory ethnographies (Knorr-Cetina 1999; Latour and Woolgar 1986; Martin 1991; Merz and Knorr-Cetina 1997; Traweek 1988). According to Sal Restivo (2005, 250), "An ethnographic approach to laboratory culture ... allows one to disentangle the intricate social machinery of fact making."

While the laboratory especially early on represented a critical site of epistemic and cultural work, analyses of scientific fact making could not neglect the sociopolitical structures that enabled this work (Kuhn 1996), financed it (Rajan 2017; Robinson 2018; Shapin 2008), accorded it authority (Dumit 2004; Fleck 1981; Foucault 1972; Gilligan 1993; Habermas 1971;

Haraway 1997; Holman, et al 2018), exported it globally (Beaulieu 2001; Ecks 2005; Kamat and Nichter 1998; Knorr-Cetina 1992), transformed it into capital (Rajan 2017; Petryna 2009; Rabinow 1997), and used it to create medicalized identities (Biehl 2005, 2007; Canguilhem 1978; Foucault 1970; Fullwiley 2011; Jain 2006; Kessler 1998; Petryna 2006; Pollock 2008; Rapp 2011). Thus, what anthropologist David Hess (2001, 237) calls the "second generation" of science and technology studies engaged "a wider field site than the laboratory ... including the political, institutional, and economic forces that govern the selection of research fields and programs."

Early ethnographic (Haraway 1989; Knorr-Cetina 1999; Traweek 1988) and ethnographically inflected scholarship (Latour and Woolgar 1986) that focused on the inside of scientific laboratories mapped the two-way translation between laboratory facts and the broader social, contextual milieu. Yet one could only map how facts move from the laboratory into the world through theoretical positing or analyses of a historical example such as that of Bruno Latour (1983) in his analysis of the pasteurization of France, or via larger inferential analyses, such as that of Michel Callon in "Some Elements of a Sociology of Translation: Domestication of the Scallops and the Fishermen of St Brieuc Bay" (1986).

Indeed, science and technology studies (STS), exemplified in the work of Latour and Callon, have had a sustained interest in the stuff of translation: the notions of boundary object (Fujimura 1992; Star and Griesemer 1999), boundary organization, trading zones (Galison 1997), and the actor network theory/approach (ANT) (Latour 2005) all attempted to get at the structures and structuring (Bowker and Star 1999) required to manage the technical and institutional factors, alliances, and allegiances necessary to transform science into facts and facts into truth. Contemporary STS work amended these early findings to consider the interrelationship between science and knowledge in relation to gender, race, sexual orientation, and disability (Benjamin 2014; S. Epstein 2005; Haraway 1990; Keller 1995; Nelson 2016; Sharp 2013; Wailoo et al. 2010; Wailoo 2017). In this research and the critical perspectives emerging with it, the theoretical intervention was not only to expose the social and political negotiations that went into the social construction of facts, but also to deconstruct the sense of those facts as having emerged naturally. In other words, these approaches critiqued the narrative that facts moved seamlessly into the world because they reflected unmediated truth.

Thus, the emergence of translational science offers a novel opportunity to explore a form of knowledge production *that begins with* the gaps between the laboratory and the world, a scientific mode explicit about aligning the "train rails" (Latour 1988, 155) necessary for the transmissibility of laboratory discoveries and their realization inside devices, agendas, and worldviews. This becomes abundantly clear in the rapid design and creation of infrastructures intended to connect research and researchers, align potential collaborators, connect capital with projects, and help integrate data across various scientific fields. TN, I argue, actually emerges from fantasies of transmissibility—fantasies made bare against the backdrop of great scientific failure. These failures included poor understanding of the neurobiology of mental illness, mushrooming evidence about the general inefficacy of newer generations of antidepressants, as well as expensive and frequent clinical-trial failures for psychopharmaceuticals. As my fieldwork shows, the response to these failures has also included the creation of novel infrastructures and architectures—buildings designed to foster collaboration, software technologies designed to durably impact research activities and collaborations, as well as semipermanent legal and commercial structures that enable translation.

## 1.6 On Neuroscience and Neuroethics

My project is a social study of neuroscience. Contemporary neuroscience is dominated by triumphalist narratives about new horizons of knowledge about the brain (Abi-Rached 2008a) and subsequently, new possibilities for cognitive, neurological, and neuropsychiatric intervention. Scholars are concerned (and rightly so) about how neuroscience might reshape social relations and society writ large via novel neurocentric notions (Dumit 2000; Martin 2007; Rose 2004). Given that these promises of neurotechnological futures rely on investments in future-oriented "returns," TN relies on speculation in ways that are similar to the speculative performances that undergird grand promises in genomics (Fortun 2008; Rajan 2006) and biotechnology (Brown 2003; Tutton 2011), and that animate sustained hope in the laboratory sciences above and beyond their actual productivity (Ioannidis 2004). There is also an interest on the part of anthropologists (Beaulieu 2001, 2003; Dumit 2004; Kirmayer and Raikhel 2009; Lakoff 2005; Lock 2002; Martin 2007; Schlosser and Ninnemann 2012; Young

1997), sociologists (Fitzgerald 2017; Joyce 2008; Rees and Rose 2004; Rose 2004; Zerubavel 1999), philosophers and historians (Abi-Rached 2008a; Changeux and Ricoeur 2002; Harrington 1989), and ethicists (Racine 2010) in the many ways neuroscience knowledge impacts society as well as conceptualizations of the self.

Yet the explosion of neuro-optimism has given rise to new scholarly concerns. A new area of inquiry called critical neuroscience (Choudhury and Slaby 2018; Kirmayer 2012) focuses on tracing and analyzing the explosive growth of the use of brain concepts in all aspects of social life, especially after the advent of fMRI. The application of, and appeals to, ambiguous neuroscience concepts and notions across all kinds of domains and scholarly disciplines such as law, behavioral economics, and addiction studies, create questions about the growth of neuroscience as a particularly prolific space of mythogenesis. Critical neuroscience approaches enable a critique of the ever-present deployment of neuro concepts bereft of critical reflection, empirical data, and/or appropriate contextualization.

Nikolas Rose has alluded to the need for empirical data about neuroscience in translation. Contrasted against the often-theoretical foci of neuroethics, he suggests that "there [have] been many debates over the implications of what could be referred to, at least over the past decade, as 'the new brain sciences' [but none] of them ha[s] been grounded 'in a sound empirical knowledge' of what is actually happening in those fields and what are the actual implications when they [the brain sciences and their applications] move 'from the laboratory to the field'" (cf. Abi-Rached 2007, 4). The absence of these data leads to Rose's caution regarding certain kinds of antineuroscience critiques—especially those against neuroscience itself or against alarmist claims about neuroscience's inevitable "harms." In their book *Neuro: The New Brain Sciences and the Management of the Mind*, Rose and coauthor Joelle M. Abi-Rached (2013) invoke a Haraway-ian indeterminism (or optimism) regarding a science that is presently unfolding. Rose and Abi-Rached call for empirical data about neuroscience and society as an obvious antidote to dataless critiques.

My project is a move in this direction—utilizing ethnographically informed analyses to look beyond or underneath the discourses created around TN toward its many materializations and constrictions. Following emerging social science research on the impacts of translational policies

(Mittra 2015), this project traces translation across several domains—from microinteractions between scientists in TN laboratories to public policy discussions at Neurotechnology Investing and Partnering Conferences—to look at how brain science is continually broken down, disseminated, reconstructed, and exported. Thus, I offer a rare possibility to map brain facts in the in-between, rather than a study of knowledge after it has left the laboratory.

## 1.7   Policies for Science and Health Innovation: Commercialization and the University

The move to use policy initiatives to inspire innovation in science and medicine and to compel economic development is at the heart of the global translational shift (Robinson 2017). Because translational science emphasizes commercialization of university research, translational science and medicine necessarily create explicit and complicated entanglements between the university and corporate stakeholders. As TN privileges commercialization-oriented neuroscience work over basic science research, this translational imperative appeals to concerns about the changing nature of the university—including its increasing bureaucratization and privatization (Apple 2005; Brown 2000; Etzkowitz et al. 2000; Geiger 2002; Readings 1997). TN exemplifies the growing entanglement between academic research and the market (Buchbinder 1993; Kahn 2012; Krimsky 2004; Shore and Wright 1999), and my research highlights the institutional influence of market-oriented approaches on knowledge production, a trend that many have mapped (Etzkowitz 2003; Haraway 1997; Krimsky 2004; Power 1999; Readings 1997; Strathern 2000).

Indeed, these concerns are not new. Thorstein Veblen voiced concerns regarding the encroachment of corporate interests on academic work at Harvard in 1918 (Veblen 1918), prefiguring the commercial transformation of universities such as MIT and Stanford after the Cold War (Etzkowitz 2007; Lowen 1997). Yet a new strain of research emerged that analyzed neoliberalism impacts on university research and the life sciences (Berman 2012; Lave, Mirowski, and Randalls 2010). My project engages with this latter literature, especially, to make sense of TN's connection to the neoliberal expansions (Strathern 2000), shaping (Hackett 2014), and contractions (Apple 2005) at the modern university.

While the university has long been at the center of political and financial shifts in the United States, these shifts also have a clear policy history.
US policies such as the 2011 America Invents Act and the 2016 21st Century Cures Act (as well as European programs such as Horizon 2020) have
sought to reshape biomedical research and regulatory approval processes.
One of the most influential factors in the commercialization of the modern
university has been the passage of the Bayh-Dole Act of 1980. Sponsored by
Senators Birch Bayh and Bob Dole, this legislation allowed US universities
to create intellectual property (IP) from research financed by public funds,
essentially crafting a route for the privatization and commercialization of
public resources. While some scholars attribute the growth of university
licensing and patenting to this legislation, others (Mowery et al. 2004) complicate this narrative. Nevertheless, the Bayh-Dole Act, alongside other legislation, coalesced with a series of shifts (Krimsky 2004; Lowen 1997; Rajan
and Leonelli 2013) that caused science at the university to be increasingly
conceivable as a space of market discovery and penetration and a sphere
where, as many scholars have noted across various contexts (Langley 2018),
private investors can support research conducive to transformation into
market value.

However, pointing only to the emergence of the entrepreneurial university and the privatization of higher education in the West does not
sufficiently contextualize the emergence of TN. My research provides the
contexts and contingencies—such as the crisis in the pharmaceutical sector
and the related reality of the global economic crisis—that lend the specificity necessary to understand the turn toward translation at modern research
universities.

## 1.8   Neoliberalism and the Financialization of Science and Medicine

My argument also engages with research on the growing import of capitalism and finance in the life sciences. Yet, in contrast to other authors, I argue
for the importance of focusing on the role of finance in order to understand
the nature and structures of translational science and medicine. As with
the biotechnology industry, the distinctive "science-business" quality of TN
creates particular challenges (Pisano 2006). For example, in the case of TN,
despite the internally acknowledged problems of drug efficacy, safety, and

reliability that plague the CNS sector, neurotechnology revenues in 2009 still reached $143.1 billion, according to one estimate.

In addition, beyond the existing profitability of CNS research lies its market potential: according to the NeuroInsights *Neurotechnology Industry Report*, brain-related illness and mental health issues impact "two billion people" globally and more than 100 million people in North America. Thus, the market size for CNS (market size being a key indicator used to make arguments for the profitability of an investment) is estimated to be $1.3 trillion (NeuroInsights 2009). With aging populations, an ever-growing categorical and diagnostic umbrella, the market size for products and services for brain-related illness is swelling. It is in the context of the massive profitability of TN that questions about the relationship between TN and health arise. In this way even the mere promise of health ushered in via TN can create access to capital.

Through its focus on biotechnology, the present project engages with the body of work that theorizes emerging scientific practice as intertwined with—and at times indistinct from—market practices. Yet much of this literature has yet to deal with neuroscience as it has with genomics and bioengineering. The "science-business" model (Pfotenhauer and Jasanoff 2017; Etzkowitz 2003; Hong and Walsh 2009; Pisano 2006) and/or scientific entrepreneurialism (Ong 2010, 5) and their connections to federal funding make the intertwining of science and business a prime example for theorizations of power, knowledge, and governmentality.

While many have theorized the interrelationship between capitalism and the life sciences, the most significant implications of changing models occur in the regime of health. And yet, as Kean Birch and David Tyfield (2013, 301) note about ongoing research on capitalism and the life sciences, much of this scholarship is inattentive to "the transformation of economic processes in modern capitalism," which provides specificity to theorizations of bioscience and the life sciences. Advancing the conversation about the interplay of capitalism and the life sciences, my project inserts critical specificity regarding financial and economic shifts that get lost in analyses of the contemporary life sciences. This project therefore contributes to scholarship focused on the emerging stakes, risks, and potentials of biotechnology as intertwined with neoliberalism (Barben 1998; Birch 2006; Cooper 2008; Dumit 2012b; Fortun 2008; Mirowski 2012; Ong 2010; Rabinow

1997, 2002; Rajan 2006; Sharp 2013; Stone 2010; Styhre 2015), especially as these interconnections implicate university knowledge practices.

Indeed, this book offers a case study about neoliberal policies as manifested in one particular set of transformations in research and innovation. In the case of TN, neoliberalism is perhaps most implicated in TN's particular stance on the role of marketability in determining what ought to be considered valuable scientific and biomedical endeavors. While I go into greater detail about this in later sections, one finds consistent suggestions by its proponents that a true realization of the translational shift requires a new way of understanding and assessing scientific work (Addison 2017; Robinson 2018). This substitution—whereby patents are better measures of scientific success and "truth" than a high-ranking journal publication—shows the way that under translational research, the market becomes understood as the ultimate arbiter of value in science and knowledge production, a defining aspect of neoliberalism's (long-standing) ideological apprehension (Lave et al. 2010; Mirowski 2012; Wailoo et al. 2010) of knowledge practices.

This study reveals that even where no transformation in patient health or new pipeline of innovations has occurred, the global biopharmaceutical R&D sector has still achieved a significant strategic goal, which is the divestment of risk, a lowering of costs, and at least ostensibly, the broadening and development of a global pipeline of potential innovation sources. Thus, I emphasize that TN's transfer of risk and its absorption of R&D cost from the private sector constitute its own *fully complete* financial outcome, even before questions of new medicines for patients or a wellspring of novel innovations arise. Even as few new medicines or innovations materialize from the economic assumptions regarding innovation that power TN's promises, it is essential to see the financial functionality of what Kean Birch (2017b) calls technoeconomic assumptions.

However, while neoliberal policy is at the heart of the case that this book outlines, my aim is not to add to the many theorizations of neoliberalism. Instead, I analyze specific, often-overlooked processes of commercialization and corporatization that are necessary to understand the complex empirical realities of TN and by extension, translational science and medicine. Here I borrow from the insights of contemporary economic anthropologists (Langley 2018; Ofstehage 2018; Pitluck et al. 2018) who seek to move beyond purely functionalist accounts to explain the expansion of financialization.

Analyzed as a political economic configuration and not merely a product of patient activism or epistemic opportunity, one sees more clearly the role of financialization (Tyfield 2011) in the emergence of TN. I combine this attention to political economy with an ethnographic sensibility necessary for understanding the many on-the-ground realities that get missed in accounts too inattentive to the unintended local and on-the-ground implications of global, national, and institutional goals, regimes, and programs aimed at fostering global innovation.

## 1.9  The Public and the Private

TN, and, I argue, translational science and medicine in general, are also born out of visions of the emancipatory power of private markets. Yet perhaps paradoxically this entails innovations in public-private partnerships that allow translational science and medicine, including TN, to offer specialized pathways to privatization while externalizing risks in the public domain. Academic translational science operates in ways that are facilitated by the provisions of the 1980 Bayh-Dole Act (as well as other policies, including the Federal Technology Transfer Act) enabling the commercialization of research carried out with public funds. For its proponents, the Bayh-Dole Act includes specific provisions that incentivize and mandate the public use or dissemination of federally funded research, to ensure a public benefit from federal science funding. Yet this shift—toward greater accommodation of private interests on the part of the state—also reflects increasing privatization, which, as many scholars have pointed out (Jameson and Miyoshi 1998), co-occurs with, and is further spurred by, downturns in markets, which was the global situation in 2006.

However, the public domain also constitutes a "hiding space" for the risks that emerge from privatization. In fact, I propose that TN rearranges risks that had been borne mainly by pharmaceutical companies so that they are now transferred to the nonprofit university. That TN invokes federal agencies, federal research dollars, and risk-laden public-private partnerships means that TN is also a case study in the public ingestion of corporate risk-taking. While risk resocialization is a known issue (Atkinson-Grosjean

2006; Avorn and Kesselheim 2011), the novel caution here may be the need to regulate optimism in light of neurotechnology's complicated realities and the financial aims of such partnerships. It is also critical to understand how the emergence of TN reflects a much larger set of private-sector transformations and financial strategies, which when uncovered, illuminate the broader logic behind the translational turn.

## 1.10   Science, Biomedicine, and Health

Lastly, this project is about the future (and futures) of medicine. The promise of translational research and medicine, according to its own rhetoric, leverages the weight of scientific research into grand, biomedical solutions to the world's pressing health challenges. Yet, to the extent that this promise entails a mode of research explicitly focused on creating biotechnologies and new potential therapeutic markets (Biehl and Petryna 2011), and promotes continuing investment opportunities despite a lack of new health outcomes (Dumit 2012a), TN lies at the intersection of R&D risk and sizable market potential. Given the massive market potential for CNS interventions, the ability to generate enormous wealth bereft of clear impacts on health, and the utility of TN to engineer solutions regarding corporate financial risks, I argue that TN should be considered primarily a form of finance.

Beyond neuroethics concerns about the potentials of TN in areas such as cognitive enhancement, there has been scholarly interest in the practical impacts of the brain sciences on local health etiologies and practices and the convergence of neuropsychiatric models with local ontologies. The tracing of these processes invokes the power of ethnographic sensibilities in tracing scientific and biomedical transformation in various registers: institutional, national (Biehl 2007; Lakoff 2005; Lock 2002; Lock and Nichter 2002), personal (Biehl 2005; Löwy 2000), and local (Biehl 2005; Das and Das 2006; Petryna 2006; Petryna, Lakoff, and Kleinman 2006). More than merely promises for the alleviation of suffering in mental health, neuroscience was supposed to usher in an era of scientific modernity positioned to impact all kinds of sciences that rely on knowledge about the brain.

Given the impact of biomedicalization on discourses around health, emerging models of health innovation have become an essential site of analysis for medical anthropologists (Biehl 2007; Ong and Collier 2005;

Petryna 2005). While I am attentive to recent work on biomedicine and biomedicalization, this project is particularly indebted to the notion of pharmaceuticalization, which as João Biehl (2013, 425) notes, compels an analytical mode that moves "beyond the unidirectional construction of patient subjectivity by medical diagnostics and treatments to account for the entanglement of multiple social forces and markets, the chemical concreteness and circulation of pharmaceuticals and illnesses, and the role of patients' agency and desires." I hope to map this "entanglement" both analytically and ethnographically in my tracing of institutions, organizations, systems, diagnoses, patients, and subjectivities as they are implicated in translation's logics and rearrangements.

Yet patients remain central in this entanglement. In what I observed during large investment conferences, which I detail in this book, I learned that patients are imagined in those spaces in very partial ways, constituting what I call an *epistemology of parts*. I ask if such ways of conceiving patients are related to the privileging of biological therapeutic solutions over other ways of imagining solutions. Might such a focus on parts also enable the tendency within translational fields to focus on distinct disease categories over more holistic or systemwide conceptualizations of patients and their afflictions? How are patients—figured as the beneficiaries of the translational shift—implicated in the shift that is translational science and medicine? In a field that sees people primarily in relation to their biological parts, what consequences emerge from a science that produces "partial subjects"?

## 2   The Histories of Translational Science and Medicine: Translation as a Political Economic Imperative

It is the responsibility of those of us involved in today's biomedical research enterprise to translate the remarkable scientific innovations we are witnessing into health gains for the nation. In order to address this imperative, we at the National Institutes of Health (NIH) asked ourselves: What novel approaches can be developed that have the potential to be truly transforming for human health?
—Zerhouni (2005)

While this book argues for adopting a political economic view of translation (drawing specifically from cultural political economy), various histories of translational science and medicine (TSM) focus on particular aspects over others. Invariably, the question arises as to why TSM emerged in particular contexts and during specific historical moments. What was behind this global emergence after 2004? Was it a product of state exasperation regarding taxpayer-funded science? Was it a crisis of patient need? Indeed, a crucial part of charting TSM requires tracing its history as a product of policy decisions and subsequent institution building. Institutional and policy-oriented histories tend to be much more attentive to the role of funding and policies in the shaping of scientific research, the role of the transformation of the university in these processes, and the role of legislative action in the shaping of science research and policy. Of course, the question of what is actually behind these institutional transformations shows how even institutional histories do not entirely elide the question of how and why TSM emerged. In the case of TN, a co-occurring history of neuroscience that created the conditions for its emergence as a translational field is key to the story.

As is laid out in the sections that follow, understanding the mysterious rise of TN and of TSM in general requires a foray into the world of global

R&D, postcrisis investor anxiety, the transformation of the university in the West especially since the Cold War, and importantly, the suturing of translational initiatives to what I have called *semipermanent commercial architectures*. These architectures confer essential durability onto translational aims, an aspect that distinguishes TSM from prior efforts at producing application-oriented work from federal funds and university research.

## 2.1   Taxpayer as Shareholder: The Emotional and Financial Politics of Bioscience

Just as my conversations with Zack illustrated the dually moral and economic character of TN, a policy history also reveals this duality and offers some insight into how TN and the translational shift emerged. Indeed, much of the history of TSM is inextricable from legislative and political performances (Robinson 2017). During the 2008 US presidential election, Republican vice presidential candidate Sarah Palin appeared before a group of supporters in Pittsburgh and addressed disability research funding. Intending to show support for special education research, she nevertheless veered off message and took on the topic of the waste of taxpayer dollars. With cameras rolling she created yet another campaign controversy (see figure 2.1). Palin informed her audience that "sometimes these dollars, they go to projects having little or nothing to do with the public good. Things like *fruit fly* research in Paris, France" (original emphasis). Palin's general concern was children with special needs. The irony—and what Palin likely did not know—is a fact that many in the media were quick to point out: the fruit fly is at the heart of the medical research focusing on treating the medical disorders she had in mind.

Palin's commentary instigated a rare public discussion of *Drosophila melanogaster* (the fruit fly) and the taxpayers' stake in publicly funded scientific research.[1] In 2008 the US economy was reeling from a massive recession and the threat of economic collapse loomed ominously over the country. Conservatives were anxious to articulate examples of government waste, especially endeavors for which immediate results could not be seen. Scientists and commentators scoffed: they pointed out how *Drosophila* research was actually *quite useful* and was an important road to cures for degenerative diseases, including Down syndrome, the disease that affected Palin's youngest son. Sentimentality about her son's condition had always been

**Figure 2.1**
Vice presidential candidate Sarah Palin in 2008 on Fox News discussing federal science research funding and the importance of finding new solutions to help children with special needs

part of her political profile. For example, she would frequently hold her son while talking about abortion and the value of life.

Palin's lament about the frivolity of fruit fly research and the need for cures was an explicitly populist indictment of ostensibly unproductive and unnecessary science research. In this view, unproductive science is knowledge for its own sake without immediate and actionable benefits to the public. Yet her larger critique also concerned the need for public money to support finding cures for diseases—all in the name of the public good.[2] Here we see a proposition about channeling federal research dollars in productive directions, rather than pursuing "useless" knowledge or knowledge for knowledge's sake, which can then be situated as elitist. The media treated Palin's populist outrage and ignorance about science as a story about Palin herself. However, for academics it reeked of a dangerous unreflective instrumentalism (Hofstadter 1963), an anti-intellectualism that regards non-application-focused knowledge as valueless (Grace 2012), suspicious, a waste of precious public resources (Wynne, Stilgoe, and Wilsdon 2005),[3] and, given missed opportunity costs to do "real science," unethical.

The complicated relationship between populist rhetoric and the democratic demands of a publicly funded science elicits a fascinating comparative history. The public value of science has been an explicit issue in Europe as far back as the Enlightenment. Accordingly, public value has long been linked to public values. There is a European history of class structures that lies behind discourse—especially of scientists—that justifies the public value of science. The discourse juxtaposes "the need for an educated citizenry" with science as the natural vocation of the intellectual and social elite and thus necessarily immune to the everyday interests of the public. Historically, instrumentalist science has had the potential for important social uses and applications, while basic science has relied on the social value of intellectualism—foregoing the requirement that one only does science that has clear direct application. These histories reveal an aristocratic bias in historical justifications for basic science (the mode of scientific investigation focused on gathering fundamental knowledge and principles): intellectual versus instrumental (practical and applied) forms of science. Here we find a historical corollary to Palin's populist critique of science set against a rising demand on the part of Congress and the National Institutes of Health for application-focused, market-oriented knowledge. Ironically, as economic anthropologist Karen Ho (2018) notes in her piece, *"Markets, Myths, and Misrecognitions: Economic Populism in the Age of Financialization and Hyperinequality,"* this populism has emerged precisely as a result of real consequences of global financialization, while at the same time obfuscating the role of global financialization in creating the economic conditions that gave rise to said populism. Nonetheless, the debates about what science ought to be doing for society hinge on varying definitions of value and subsequently diverse methodologies for assessing its societal benefit (Wynne, Stilgoe, and Wilsdon 2005).

Palin's fruit fly controversy raises several issues. First, the laboratory fruit fly, much like the laboratory rat, has contributed in innumerable ways to mapping disease models in animals and has led to important research and medical therapies. Much of the public backlash regarding her statements focused on the potential contribution of *Drosophila* to life-saving cures, which, perhaps ironically, shows how the lab rat and *Drosophila* have come to symbolize biomedical and translational *possibility* (Davies 2012), and accordingly, the future of health.[4]

Second, in the context of the 2008 financial crisis, Palin gave voice to complaints about two interrelated problems that impinge on federal funding for science and medicine. According to one narrative used to explain the emergence of TSM, a rise in populist sentiment among large sectors of the public questioned federal spending in general, and members of Congress expressed concern that the enormous federal investment in science has yielded little. The populist narrative expresses social anxiety regarding hard-earned money being sucked into the black hole of basic science, positioned against translational or applied science that develops clear application-oriented scientific research. Especially in the case of health and biomedicine, translational science is positioned as a necessary precursor to global and other large-scale innovations in health.

Thus, a significant part of the history of TSM intersects with long-standing debates about public investments in research and the demands that such investments produce results benefiting the larger public. Concerns about government waste, poor government accountability, rising higher education costs, as well as questions about the value of public education and publicly funded university research, all became part of the legislative warrant associated with application-oriented demands made by the state. However, this is only part of the story that helps to explain the rise and sudden emergence of TSM.

## 2.2 A History of Translational Science and Medicine

*Translational science, translational medicine, clinical translation,* and *translational research* are all terms representing a common set of overlapping concepts that coalesce around a central notion: the complete redesign of biomedical research to accelerate the process whereby science is turned into applications and especially products. While this book focuses on neuroscience, there are a variety of translational fields, including translational biology, translational oncology, and translational bioinformatics, among others. Thus, *translation* is used to refer to a wide set of activities, projects, and initiatives that stem from this abstract notion.[5]

The birth of the explicit translational pipeline at the NIH began between 2003 and 2005, though state-sponsored research aimed at translation has had a long history. Canadian oncology research in the 1990s is a very important flashpoint in this history, especially in the formation of translation-oriented research modes (Cambrosio et al. 2006). However, interestingly, between 2005 and 2009, the financial crisis coalesced with new significance around risk and investment—especially in the commercial biosciences. Thus, the growth of translational science and its *institutionalization* at universities in the United States occurred at a particular time, during which pharmaceutical companies were scaling back neuroscience investments, access to global capital was becoming more constricted, and political narratives—such as those from Palin—heightened anxieties about the future of investments in science and medicine.

One challenge in tracing the rise of TSM is demarcating its conceptual emergence as a *zeitgeist*, an ethos that gained traction at a particular moment and in a particular context, versus mapping its *institutionalization* via discrete practices, policies, and actions. This challenge matters for how one writes and understands the history of "translation." The brief history below—again, only part of the story—contextualizes the sociopolitical environment that surrounds the translational turn in science and medicine.

In October 2005, Elias A. Zerhouni, then director of the National Institutes of Health, provided an interview and published an influential article, "Translational and Clinical Science: Time for a New Vision," in the *New England Journal of Medicine*. According to Zerhouni, the NIH started in 2002–2003 to create a series of collaborative discussions with scientists, clinicians, university administrators, entrepreneurs, patients, and other stakeholders. Out of this series of dialogues, the NIH produced the NIH Roadmap for Medical Research. From this roadmap, the NIH launched a new program in 2005 to fund Clinical and Translational Science Awards (CTSAs) to US universities. CTSA was the funding program that gave institutional birth to university translational science and translational medicine on a large scale. As Zerhouni (2005, 1621–1623) put it, "We now aim to stimulate the development of a brighter vision for translational and clinical research, to ensure that these disciplines remain powerful engines of creativity. We offer the opportunity for change to those who share a vision and commitment to innovation and experimentation."

Suddenly, starting in 2006–2007, universities across the US began to create new centers whose names contained the term *translational*. One finds a sudden sprouting not only of translational centers on university campuses, but of doctoral training programs, new buildings, new research projects, and more importantly, new collaborations—all stemming from this specific program.[6]

For example, the University of Alabama formed the UAB Center for Clinical and Translational Science and UCLA created the UCLA Clinical and Translational Science Institute. These awards also facilitated the creation of multi-institution centers such as the Georgetown-Howard Universities Center for Clinical and Translational Science in Washington, D.C., and the Atlanta Clinical and Translational Science Institute (Atlanta-CTSI), which is a network connecting Emory University, Morehouse School of Medicine, Georgia Institute of Technology, as well as Children's Healthcare of Atlanta. These new centers produced new constituent translational programs—of which TN is an example. These programs have created entirely novel configurations at universities: new departments, new PhD programs and funding schemes, new buildings, and even the renaming of buildings. NIH's persuasive power can be seen at Tufts University in Massachusetts, where funding from the NIH meant rearranging entire departments and centers to fit new funding schemes attached to translational science.

This NIH program created significant change in the landscape of US university research: in 2005 alone, 60 medical research institutions across 30 states and Washington, D.C., were created with one fell funding swoop. Because the NIH is the largest government funding body of health research in the world (Ioannidis 2004), this shift represents a massive sea change regarding the institutionalization of visions for health.

While the NIH catalyzed TSM in the West, the globalization of TSM had already begun. Shanghai Jiatong University spearheaded the formalization of TSM in China, creating the Med-X Institute of Translational Research by 2007 and the Academy of Translational Medicine in 2011. By 2014, China had more than 50 TSM institutions (Dai, Yang, and Gan 2013), with stated intentions to build another five institutions in 2014. By 2010, a group of European countries created the European Advanced Translational Research Infrastructure in Medicine (EATRIS) with the goal of developing TSM centers (Shahzad et al. 2011). In the UK, the Wellcome Trust Scottish Translational Medicine and Therapeutics Initiative connected four academic medical centers in Scotland, while Cambridge and Oxford Universities created TSM centers soon thereafter. TSM projects emerged in the Netherlands as well as Austria, Australia, South Korea, and India. By 2016, TSM had become an inextricable part of a global biomedical modernity with hundreds of centers focused on it.

Yet, even the NIH's funding of individual translational science centers was not enough. In 2011–2012 the NIH poured $575 million into developing the new National Center for Advancing Translational Sciences (NCATS) to "transform the translational science process so that new treatments and cures for disease can be delivered to patients faster."[7] According to its mission, "The Center strives to develop innovations to reduce, remove or bypass costly and time-consuming bottlenecks in the translational research pipeline in an effort to speed the delivery of new drugs, diagnostics and medical devices to patients." In essence, and with an enormous effort, the NIH created its national research center in ways that paralleled what it was funding other institutions to do. A priority of the NIH director, the NIH's FY 2013 budget request for NCATS totaled $639 million (out of a total request of $31 billion).[8] In the same year, the NIH allocated $110 million to fund up to 18 CTSAs (see figure 2.2). The NIH's budget request for NCATS for 2017 was $685.417 million.

The creation of NCATS was momentous not only because it involved the creation of an entirely new national center, but because it represented a US intervention in the market via the direct licensing of its discoveries to private companies. Essentially, the NIH created its own entity to accelerate the instrumentalization of scientific knowledge: on the homepage of the NCATS website the program advertises "NCATS Inventions Available for Licensing." There, an entire list of inventions from NIH laboratories is available for licensing. Through NCATS, the federal government can directly partner with firms interested in investing in the development of its intellectual property (IP) as well as ensure that NIH investments in science and medicine result in therapeutics. According to NCATS, "Inventions made by NCATS employees are owned by the federal government. Patenting and licensing are used to help ensure the technology is fully developed, commercialized and advancing public health."[9]

Here we see a critical suturing of narratives of translational imperatives to benefits in the form of public health.[10] Importantly, it also means that the state had become more explicit and intentional about entering the business of creating assets in the form of IP. Both NCATS's licensing of its own innovations and the move to push licensing of federally funded institutions'

**Figure 2.2**
A brochure from the Translational Center at Tufts University, taking on much of the language from the NIH initiative

innovations reflect the impact of legislation such as the Bayh-Dole Act and the Stevenson-Wydler Technology Innovation Act (both signed into law in 1980), the 1986 Federal Technology Transfer Act (alongside IRS revenue procedures), as well as more recent legislation, such as the 2011 America Invents Act. That NCATS initiatives facilitated direct partnering activities indicates the sheer fervor at the national level to turn federally funded science into market value.

Indeed, the translational shift serves a clear market need of readying a supply chain of IP around which to mobilize financial investment. It also addresses a central anxiety about how to ensure that the massive funding and energy expended on national science are productive—for society, for the market, for potential patients, and for the public. Not only does NCATS compete directly with the universities that it also funds, it also represents a remedy on the part of the state designed to ensure the translation of federally funded research into commercial, medical, and patient value.

Thus, Sarah Palin's seemingly misguided comment on the absurdity of fruit fly research was actually in step with a larger theme. According to some, the slowdown in new patents and rising healthcare costs tracked dwindling congressional patience: "Congressional leaders, policy-makers, and the public at large are increasingly concerned that the scientific discoveries of the past are failing to be translated into tangible benefits to public health. The response has been a series of initiatives making translational research a priority" (Garcia-Rill 2012, 1).

Yet not all translational science is created equal. TN is both one of the more promising endeavors and one of the most difficult. The complexity of the brain and the particular difficulty of understanding brain disorders (especially in relation to mental health)[11] have made it quite risky in terms of continued development for pharmaceutical companies and investors (Cutler and Riordan 2011). At the same time, TN also represents an enormously lucrative global market opportunity.

## 2.3   Translational What? Defining Translational Neuroscience

In the same controversial speech, Palin, if unintentionally, took a detour through TN. She said, "Early identification of a cognitive or other disorder, especially autism, can make a life-changing difference." While the early 2000s witnessed a movement among government science policymakers, university administrators, venture capitalists, and entrepreneurial academic scientists to reshape the goals and methods of biomedical science, *translational neuroscience,* in particular, promised to channel the breakthroughs of brain science into life-saving and life-enhancing diagnostics and cures. However, what exactly is TN?

The issue of identification or diagnosis of brain-related disorders, often called *diagnostics,* is one key focus for TN, the term that caught on culturally in the United States between 2005 and 2010. The difference between TN and basic neuroscience is complex in its implications and while TN centers describe their missions somewhat differently, they do share a conceptual core: all break down disciplinary boundaries, seeking to merge form and content; they achieve this through the promotion of cross-disciplinary work, often exemplified in the building of new centers designed to create collaboration across sectors. These institutions engage actors from outside the scientific community—venture capitalists, politicians, pharmaceutical companies—and they eschew knowledge for knowledge's sake in favor of propelling knowledge from "bench to bedside." TN is, therefore, a field of inquiry that attempts to take the insights and knowledge of basic neuroscience and clinical research and convert them into practical applications, preferably applications that can be turned into commodities.[12]

In general, TN centers are intently focused on the processes of knowledge creation as well as on its output: medical therapeutics, diagnostics, and even knowledge that improves health outcomes. In the US, Duke University and the University of California at San Diego (UCSD) both describe their centers in ways that focus on different aspects of TN, but the University of Nebraska Medical Center description sums up the enterprise well:

> The Center for Integrative and Translational Neuroscience is the link between basic sciences discoveries and translational implementation in the clinic. It will take the fruits of neurodegenerative research from the bench to the bedside. The center brings together basic research faculty with clinicians in neurology, psychiatry, pediatrics, radiology, medicine and surgery. This joint effort will move new discoveries rapidly out of the lab and into the clinic. Neuroscience is now at the stage where we can move laboratory findings up through molecular, cell culture and animal models to people to alleviate the burden of neurodegenerative illnesses as soon as possible. ... Advances in technology allow us to diagnose diseases of the brain earlier and more exactly. It is at these early stages that treatments would likely be more effective.[13]

Referencing the development of diagnostics, this description clearly designates TN as science done in the service of predetermined applications. Undoubtedly, however, this approach toward thinking *with diagnostics* creates questions about the implications of TN for scientific thinking and TN as an intervention in health. For example, how might presumptions of diagnostics as an output impact the design of research projects and the assessment of said projects? This question relates to what I have called *translational thinking* on the part of scientists, especially neurobiologists, research university administrators, and attendant stakeholders, as an explicit output of the translational paradigm. Translational thinking encompasses the outcome (or product) of experimental research, the presumptions on which researchers construct research agendas, and experimental decisions that emerge from such presumptions. Presumptions of a translational endpoint also impact decisions about which questions to pose and which data to cull.

Two prominent ideas shared across various stakeholder definitions of TN include the notion of *discovery* and the concept summed up in the often-used colloquialism, "from *bench to bedside.*" Translation is thus about turning science into medicine—and includes at least rhetorically a revamping of processes and systems to help improve or accelerate this process. However, given the enormity of all that exists between bench and bedside, translational science and medicine are quite amorphous and as such prone to definitional diffusion, even among insiders. The essential commonality among definitions is a productive end value: a new drug or diagnostic or even new therapeutic models, as made explicit in the imperative from UCSD's TN Institute: "Develop novel therapeutic technologies." Such productive "end values" can also include clinically relevant psychopathological knowledge—information that one can act on in important ways.

Yet TN is also very much about the brain. Advances in knowledge of how the brain works form one of the intellectual cores that animate the field (FitzGerald 2005; Gomez 2006; Rose 2004). The explosion of brain science discoveries since the 1990s (Abi-Rached 2008a; Garcia-Rill 2012; Vrecko 2010) and the extensive use of functional magnetic resonance imaging (fMRI) (Dumit 2000; Joyce 2008; Rapp 2011) generated a moment in time that the Library of Congress and the National Institute of Mental Health (n.d.) dubbed the "decade of the brain," signaling not simply the growth of neuroscience, but importantly the possibilities for what that knowledge could create. Accordingly, translation offers the oportunity to "downstream" (Rajan and Leonelli 2013) the wealth of (in this case) neuroscience knowledge while it also upstreams pharmacological and biotechnological envisioning on the part of neuroscientists conducting research. According to the more optimistic and disinterested narrative, the sheer accumulation of basic research created conditions for the takeoff of TN and the proliferation of new medications, devices, and diagnostics.

TN may also become a means by which an individual disorder is connected to brain science. Take for example the fact that the clinical practice of mental health has been wholly transformed via neuroscience (Rose and Abi-Rached 2013). It is the case, as scholars have noted, that "diagnosis is now thought to be most accurate when it can link symptoms to anomalies" (Rose 2004, 408). In this sense, it is possible to view TN as a mechanism that produces what some scholars have termed "patients-in-waiting" (Dumit and Greenslit 2005) by matching disorders with biological material and creating more avenues through which subjects become entangled with diagnoses, diagnostic criteria, diagnostic testing, and treatment. This is especially the case for something called the *mechanism of action*. This term refers to the biochemical means through which a pharmacological effect is achieved (Pritchard et al. 2013). This is all important in the larger bio-marker search that is part of understanding complex brain-based disorders (Lock 2007), a primary means of pharmacological meaning making around both the biological conceptualization of a disease and the rendering of a disease as an object around which to construct biotechnologies and narratives about the biotechnologies (especially for regulatory purposes).

"Biological material" refers to the panoply of biological "objects" and their subsequent meanings that become essential parts of biological explanations and understandings. In other words, biological material is the material accoutrement such as genes, cells, blood, and facts that lend reality and mobility to scientific and medical discourses. This sentiment of matching disorders with biological material is about how TN, in its appropriation and relocation of biological facts, subsequently enables said facts to be enveloped within larger narratives of disease. In the constructivist view of how diseases attain social reality, TN is perfectly positioned to produce facts and research that surround particular diseases, and even when those facts fail to translate neatly into patients' experiences or symptoms, the very possibility of relevant scientific facts can be marshaled to provide reality to disorders. It is this very capacity for TN to construct its afflictions in some sense that makes it compatible with Margaret Lock and Vinh-Kim Nguyen's ideas regarding the technological construction of disease via biomedicine. As TN becomes a means by which to "objectify" the brain and associated disorders, the body and its conditions become *problematizable*.

## 2.4 Translational Neuroscience as an Opportunity via Scientific Excess

Emic definitions of TN often focus on the therapeutic opportunities enabled precisely because of the mountain of new knowledge emerging from neuroscience. The cognitive revolution has led to a bevy of scientific research and findings about the brain (Gardner 1985). According to the director of the University of Nebraska's Center for Integrative and TN, Howard S. Fox, "Neuroscience is now at the stage where we can move laboratory findings up through molecular, cell culture and animal models to people."[14] Official definitions (on websites, in brochures, in requests for proposals (RFPs) for federal funding, as well as in market analyses focused on the neurotechnology sector) describe TN in triumphalist terms—opportunities engendered through neuroscience's excess. In other words, it is the perception of an abundance of new knowledge about the brain or of neuroscience's rapacious growth that informs the view of neuroscience as an opportunity.

Here neuroscience is constituted as a treasure trove. This narrative is pervasive in the literature from investment groups and market speculators such as those attached to the local neurotechnology community in Northern California. Since the investment literature represented the brain as a market frontier, it was not much of a stretch of the imagination for a pharmaceutical executive at a Neurotechnology Investing and Partnering Conference—an event pairing scientists and investors—to hype TN as "mining for gold." In this sense, translational science is both a goldmine and a "mine." It is both an opportunity and a risky endeavor, made available by a wealth of neuroscience knowledge.

Paralleling the discovery-wealth narratives found in TN initiatives, many narratives treat the history of TSM more generally as a nearly inevitable outgrowth of novel scientific achievements. On the heels of the Human Genome Project, for example, TSM emerged, per this explanation, as a response to an abundance of scientific discovery and a moment of unparalleled potential. Claims of a mismatch—between the rapid pace of scientific discovery and the sluggish pace of applications therefrom—accompany explanations for the *why* of TSM. Thus, in the NIH Roadmap (2003), out of which many TSM institutions emerged, as well as the policy narratives justifying the need for TSM across the West, one sees the language of scientific wealth and possibility, of unbridled new opportunities to turn the nation's wealth of knowledge into life-changing discoveries. In this treatment, TSM emerged in response to the march of scientific progress.

## 2.5   Translational Neuroscience as a Solution Born from Failures: A Crisis for Science and Patients

There are other histories, however, that treat TSM as the product of patient-driven activism. Here, TSM's goal of focusing scientific activity toward the alleviation of the world's suffering utilizes a more populist narrative to explain TSM's development and spread. Important scholarship about patient collectives that emerged in response to the AIDS crisis in the 1980s (Epstein 2005) and organization around cancer (Wailoo 2017) and other areas shows the way patient communities helped demand that scientific agendas become more responsive to the needs of patients. For some commenters, economic histories of TSM focus too much on economic issues (Aarden et al., forthcoming) and neglect more comprehensive discursive, sociopolitical structures that gave rise to TSM. This history also dovetails with the rise of foundations and what some have called "philanthrocapitalism" or the philanthropic state (McGoey 2016), referring to the emergence and growth of powerful foundations that have pushed research in very specific disease areas such as cancer, HIV/AIDS, and autism.

This narrative of patient need—which I also discuss in a later chapter—has, ironically, also become part of a separate set of discourses that position TSM as a means to recoup the enormous investment in scientific research over the last 50 years (Chubb 2012). In this latter narrative, one still finds the view that TSM emerged to reorient the mass of public science funding toward more aims serving the public good—both regarding medical cures and especially in terms of economic development. These twin explanations—about the need to fund research that better helps patients and the need to improve the dreadfully low yield of new discoveries from publicly funded science to restart local economies—both position the rise of TSM as a response to scientific failure.

Similarly, the colloquialism "from bench to bedside," referring to efforts to bridge the gap between laboratory discoveries (the lab bench) and the patient's bedside, appears in expert and official definitions of translational science. Translation is represented as a means of fixing barriers that prevent the transformation of knowledge into practical results. The NIH, for example, uses the language of overcoming barriers. Thus for many—legislators, wary investors, frustrated academic scientists—translational science largely emerges from productive failure, particularly the failure to produce new drugs or safe, effective therapies despite a mountain of federal research (with the NIH spending $33 billion on medical research and an estimated $5.5 billion in neuroscience across all departments in 2016[15]). A primary factor in this failure is called the "pipeline problem": a slowing of new effective drugs.

In a 2005 opinion piece in the journal *Nature Reviews*, Garret FitzGerald, then a leading scientist at the Institute for Translational Medicine and

Therapeutics at the University of Pennsylvania, proposed the institutionalization of "translational medicine and therapeutics" as a field. In the
article, he makes an argument for the financial opportunity here and why
this new discipline would do better than previous failed attempts to create application-focused solutions. His remarks reveal some of the problems
that translational science seeks to remedy:

> Perhaps it is time to found a new discipline of translational medicine and thera-
> peutics. ... Why establish a new discipline? First, a new "brand" creates excitement
> and enhances visibility. ... Second, previous labels for effort in this area have not
> caught on. Divisions of Molecular and Experimental Medicine have been tried,
> but have not flourished in AMCs [academic medical centers]. Moreover, the few
> examples have tended to favor basic rather than translational research and have
> lacked an emphasis on therapeutics. ... [Translational medicine and therapeu-
> tics offer] a much broader fiscal opportunity to AMCs—namely, the increasingly
> efficient commercialization of their intellectual property. (FitzGerald 2005, 817)

FitzGerald then offers his own characterization of translational medicine
and therapeutics: "The term 'translational' medicine and therapeutics captures the momentum of the current focus on the translational process,
implies the projection of basic science into the domain of clinical medicine
and emphasizes the object: the discovery of new and safer therapeutic entities" (FitzGerald 2005, 817).

However, it is important to note that although there was a pipeline problem, revenues had not completely dried up for pharmaceutical companies.
In fact, these companies found strategies to maintain profits even with a
dearth of new drugs during the height of the industry's pipeline crisis. As
FitzGerald (2005, 815) notes, "A high profit margin in the United States has
shielded drug development from the usual economic pressures that would
ordinarily drive reform." For him and others (Applbaum 2009c), pharmaceutical companies had been using a panoply of innovations to sustain
profitability despite a lack of innovative new products.

There is yet another place where translational science's object-oriented
emphasis (its failed productivity) appears. Scientific research articles typically conclude with a claim about clinical applications, even as the failure
of this research to turn into applications was widely known. In 2003, several
researchers traced the frequency with which research findings were actually
translated into clinical applications. They screened research reports from
1979 to 1983 in six prestigious basic science journals and found that despite

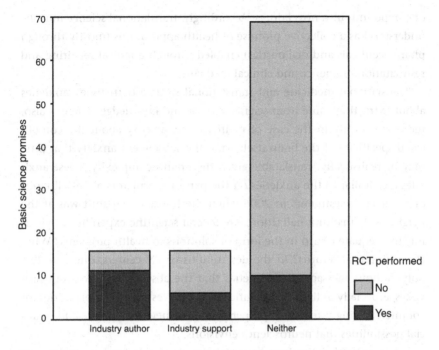

**Figure 2.3**
This table reflects a meta-analysis of research publications compared to the number of actual "applications" as constituted by the creation of a randomized controlled trial (Ioannidis 2004).

the ritualistic promise of application or the possibility thereof, few findings (see figure 2.3) actually turned into clinical applications, despite inspiring randomized controlled trials. Only one, according to the authors, materialized as a solution with major medical impact (Contopoulos-Ioannidis et al. 2003; Ioannidis 2004). Since that time, many have mapped a plethora of other problems facing biomedical research globally, including and especially the problem of reproducibility for many experimental results (Baker 2016).

Therefore, it is the *failure* to translate research in the life sciences into safe and effective pharmaceuticals that constitutes the primary problem that translational science seeks to solve. In fact, the growth of translational science as a field emerges out of a collective sigh of exasperation, one shared by federal agencies that fund US scientific and medical research and pharmaceutical companies struggling to turn their formidable libraries of compounds into novel drugs. Accordingly, translational science must be understood as a collective promise of health applications (mostly through pharmaceuticals and diagnostics) enabled through a radical repairing and resuscitation of science and clinical medicine.

Translational medicine and translational science both index anxieties about extracting value from scientific labor and knowledge. There is also, more specifically in the case of neuroscience, anxiety about the complicated specificity of the brain itself—and the subsequent anxiety that little may be realistically translatable given the brain's complexity. These anxieties are similar to the anxieties on the part of speculators about all kinds of financial investments in 2006 when the translational shift was at the height of its institutionalization: Are federal scientific expenditures yielding an adequate return in the form of solutions to health problems? What about financial returns? In the descriptions of TN centers, one detects a polyphonous message: the anxieties that the absence of new drugs provokes, especially in light of dramatic levels of investment, opportunities for commercializing intellectual property, and the new therapeutic and financial possibilities that neuroscience envisions.

While official definitions of translational science tend to underscore opportunities and innovations in science, private conversations have often yielded more anxiety-influenced definitions. For example, when I asked Zack Lynch about TN, his definition focused on the "valley of death" (Vanderheiden and Tobias 2000) as well as its necessarily pharmacological end and starting points. For Zack, translational science must "get a compound out of an academic lab, acquire initial funding, put together a small business team because you are a professor focused on translating neuroscience, getting that object out of the marketplace of ideas. That's the valley of death." On his take, TN is about rehabilitating a pipeline of underleveraged assets or remedying the barriers to commercialization. Thus, his definition is marked by anxiety and obstruction.

This definitional tendency punctuated interviews with Neurotechnology Industry Organization members and was apparent in the language used by Silicon Valley venture capitalists looking for early-stage investment opportunities. Those definitions begin with a lack of new, effective drugs as the primary problem that translational science and medicine are to solve. They focus on the lack of new compounds for mental health compared to the enormous investment in both sciences (both federal and private).[16]

Of course, the real problem of generating new and efficacious compounds may actually be the presumption of translatability that gets attached to much of the relevant scientific work. As discovered by many anthropologists, sociologists, and historians of science, laboratory work is a highly defined and prescribed context for model making (Knorr-Cetina 1992; Latour and Woolgar 1986; Traweek 1988), rendered powerful through discourses of translatability and made legible only by bringing the laboratory to the world (Latour 1983).

*Presumptions* of translatability have thus produced a distinct problem. An essay published in 2011 in the *Journal for Clinical Studies* articulates this stance for neuroscience: "Unfortunately, the large amount of information gleaned from recent basic science innovations has had very little clinical relevance and, despite newly acquired knowledge gained on an almost daily basis, the past few years of CNS drug development have been characterized by relative stagnation" (Cutler and Riordan 2011, 15). Yet the growth of funding in university neuroscience based on assumptions of translatability continues as part of the translational imaginary (Robinson, in preparation).

## 2.6  The Novelty of Translational Science and Medicine

In thinking about the history of TSM, a question emerges about its novelty. Despite a multitude of different definitions and usages, the term refers to a vast array of activities that coalesce around the central aim (and metaphor) of "translation." It is thus an epistemic project because of how it moves to integrate various disciplinary knowledges, methods, and approaches, as well as a metaepistemic one because of its focus on the redesign of research practices and knowledge systems themselves. With more than 400 new institutions dedicated to translational research and billions of dollars of investment across international agencies, the rise of TSM (including TN) has been nothing short of tectonic. Few structural shifts have had as dramatic an effect on the landscape of contemporary biomedical research institutions across the West.

Yet alongside its explosive growth and expansion lies an often-observed curiosity: TSM does not appear to be at all new. For many critics, the aims of TSM are perhaps, in the words of philosopher Miriam Solomon (2015, 175), "as old as scientific medicine." Absent novel theories or scientific methods, breathless critics wave TSM off as one among a milieu of seemingly perennial fads in biomedical "innovation." But I argue that these dismissals rely on a faulty understanding of TSM in its current incarnation. The recent emergence and formalization of TSM represent an expressly economic transformation. As the product of a specific intermingling of stakeholder economic interests, key industry reconfigurations during and after the recent global financial crisis, as well as a new infrastructure to facilitate commercialization (what I call commercial architectures), it reflects a coordinated move to redesign biomedical research according to specific financialized models. To be clear, I argue that these commercial architectures are a primary means by which TSM is different from previous incarnations. The sense of TSM as an "old trick" with a new brand, or a "New Light through an Old Window," to reference historian Alison Kraft's (2013) writing on the issue, may be a consequence of analytical modes that conceptualize scientific enterprises in primarily *epistemic terms*.

To show why a primarily epistemic framing is not the appropriate way to make sense of TN and TSM, the next chapter traces the economic contexts and financial pressures that surrounded the translational shift between 2004 and 2014. Therein, I show that TSM is not what it seems and that it must be understood—in a broader, integrated analysis—as the product of a particular political economic history and as a manifestation of an increasingly financialized model of biomedical innovation.

# 3   Science as Finance: The Financialization of Translational Science and Medicine

A RISK is a potential for a LOSS.
The LOSS is the realization of that negative potential.
A RISK is running across a busy street blindfolded.
A LOSS is getting hit by a car while doing that.
—Riskviews (2011)

In this chapter, I propose that the emergence of TN was not merely about "innovation" per the glittering narratives used to fund and justify it. In fact, TN must be understood in relation to a strategy undertaken by biopharmaceutical companies, which outsourced the riskiest parts of early-stage neuroscience R&D to universities. Consequently, translational partnerships between private industry and academic medical centers essentially turned university research projects into development arms for industry. Yet, a more in-depth look unearths how translational systems sought to enable optimal legibility between university science and the needs of the market. Since the translational shift enabled shareholder value creation through lowering R&D costs and outsourcing of R&D risk, it may be better to analyze TN as a *form of finance* rather than merely a new form of science or medicine.

Through immersion in the worlds of neurotechnology and life science investing, I learned about hidden worlds of shareholder anxiety, sectorwide divestment from research for Alzheimer's and depression, the reimagining of university science as an underleveraged asset, and a swelling private-sector opportunity represented in the translational turn. Across it all, one finds a story of science, innovation, and financialization.

## 3.1   Innovation and Risk

"We had always been focused on complementing our investments of capital into our portfolio companies with expertise, connections, and other support services to ease commercialization of the IP," Michael Hall, a partner at a prominent venture capital firm in Silicon Valley, told me in 2010. His firm has always been involved in translation, he argued. Michael shrugged his shoulders and peered out his office's large picture windows at the Presidio. The Presidio is a massive and beautiful green park area in San Francisco that was opened up for private development and now houses a stock of residences and offices. "It was interesting to see that the government was now pushing this too," he said, referencing the push by the NIH toward translational research at universities. Michael commented that the government was focused on placing commercial demands on their research portfolio in ways that he had never seen in his "many years in venture capital." He continued, "We'd always had guys who came to us looking for money, and they had an idea, and they had started a small biotech thing on the side, but now I see really interesting things happening in this space [the biotech investing space] that's bringing even old investors out of the woodwork" (Interview 10, December 12, 2010).

Michael was referring to the reluctance of some investors—especially in the San Francisco Bay Area, where there are lots of capricious investors—to continue investing in biotechnology startups. Michael then said something illustrative about biopharmaceutical strategies: "You know, Pharma doesn't care. They are going to make their money. They have already counted cash by selling shitty new copycats or some effective marketing platform, but the little guys, we really care about the science. We can't gimmick ourselves out of the problem."

Michael worked at a large pharmaceutical company before jumping to the world of venture capital. A former vice president for business development, he was brought on staff at his current firm precisely because of his connections to startups and because of his understanding of industry trends. For venture capital firms (often called VCs) that focus on life science investments, there was and continues to be a revolving door with Big Pharma. Michael's statement, however, underscored what could be thought of as different investment strategies, and thus differing conceptualizations of risk. This was a valuable insight: Big Pharma had the resources to leverage marketing (Applbaum 2009c) and other creative strategies to withstand psychopharmaceuticals' history of poor efficacy and to create new revenue streams in light of a lack of new drugs. For Michael, who considered his multimillion-dollar VC one of the "little guys," the lack of innovation from neurotechnology startups was a huge problem, and no amount of marketing or business strategizing could solve it. Thus, for him, good science had to be the way out of the investment rut.

This conversation with Michael provided a glimpse into the industry-related terrain on which TN and translational research in general are framed. For pharmaceutical companies, efficacy and new compounds continued to be a problem (Ninnemann 2012). However, because biopharmaceutical companies created new disease markets (Dumit 2012a; Rajan 2017) and new profit strategies using non-science-related strategies (e.g., marketing pushes, rebranding old drugs, expanding diagnostic criteria, or making minor molecular changes to drugs for patenting purposes) (Applbaum 2009c; Moncrieff 2006; Pisano 2006), they achieved a critical insulation from the realities of problems with products and patients. Here, the Ponzi-like quality (Mirowski 2006) of "innovation strategies" that don't rely on actual innovations in science and medicine come into view. Nonetheless, all stakeholders—from pharmaceutical companies to investors—agreed that good science was good for business. For this reason, if for no other, translational research was thought of as a good thing in its ostensible focus on doing "good science" (Wehling 2015).

One could argue that Michael's claim about VC firms "caring about the science" was rather partial since small VC-backed biotechnology companies also find means of "insulation"—which is to say that they often benefit and profit from hyping and other strategies of value creation that extend beyond mere "naked" scientific innovation (Brown 2003; Cooper 2008), just as pharmaceutical companies use marketing strategies to drive up revenue despite a lack of new compounds (Applbaum 2009c; FitzGerald 2005). One key example lay in the strategies used to "overvalue" startup firms even where there is little novel technology or IP (Sunder Rajan 2012).

Nevertheless, conversations with Michael and other investors indicated a paramount need to look at how investors and Big Pharma created risk strategies given the decoupling of market value from scientific value. Translational science and medicine were about creating and managing market value in a context of significant scientific uncertainty.

## 3.2  Partnering and Licensing in Neurotech

On the last day of the 2010 Neurotechnology Investing and Partnering Conference that I attended in Boston, there was a morning panel that was sparsely attended. The event was a premier event that attracted academic scientists, scientist-entrepreneurs, investors, pharmaceutical executives, and biotech startups looking for funding. It also functioned as an annual "congress" of sorts where sector issues were announced and partnerships emerged. This panel was devoted to "Partnering and Licensing in Neurotech," but its focus belied a sense of crisis. In the panel description, licensing and partnering were clearly elucidated as avenues through which to remedy pipeline problems, but the description focused on questions regarding emerging strategies to remedy this problem:

> Big Pharma, biotech, and Medtech are increasingly looking to smaller firms to provide innovative product candidates to fill pipelines. What are current trends in strategic alliances and partnering? Is there a new model for partnerships emerging? Is Big Pharma looking to earlier stage product candidates for partnering and in-licensing? Can major medical device companies continue with their high rate of acquisitions?

Tasked with answering these questions was the late Jeffrey Nye, a tall, youthful man with rimless glasses and a BlackBerry that seemed never to leave his hands. Even during the panel in which he participated, he checked it constantly. A PhD, MD, and Harvard graduate, Jeffrey exuded confidence, which even spilled into his informal interactions with the other panelists. He attended the event as the vice president and head of External Innovation for Neuroscience at Johnson & Johnson, a large multinational medical device, pharmaceutical, and consumer-goods company. On this panel dedicated to internal and external research and development, Jeffrey explained how public and private partnerships allowed Johnson & Johnson to lower the cost of bringing a drug to market. Then he pitched:

> We're still looking for partners! ... We're partnering largely with academics—recognizing that [other] Pharma companies have abandoned this space [including psychiatry, antidepressants, Alzheimer's] and this offers us an opportunity to get involved with innovation at its earliest stage and not start the race when publications start appearing and start the race with everyone else. We have created partnerships with Johns Hopkins, and our attempt is to capture this innovation at its earliest stage. ... So we have external discovery engines to afford us

the opportunity to offload the overflow from our academic collaborations. In the past, we signed a deal with Vanderbilt and … we looked at external development partnerships, and we're looking to capture the innovation of academics, and we're looking to increase our partnerships. One of our objectives is to increase early stage partnering [so that] we can get the biotech companies to do the experiments that we want. We're able to come in at a lower cost.

There are several telling parts to Jeffrey's statement. First, he praised the opportunities that early-stage investing offered, yet he also situated this new opportunity as part of a new Johnson & Johnson strategy—one focused on *early-stage innovation*, the stage at which pharmaceutical companies partner with scientists in university laboratories, rather than a standard partnership in which pharmaceutical companies would license a late-stage intervention or piece of IP from a university's licensing department. This distinction demarcates how corporate partnerships may affect early-stage or upstream knowledge practices at the university—even in the realm of the hallowed scientific laboratory.

A second important point is how this early-stage partnership strategy constitutes an *investment* strategy. That is, university-industry partnerships around TSM are not just about science or mere innovations in health, but are about an early-stage investment model. Such approaches benefit investors both because of the relatively "low price" for them since such initiatives are not far along in terms of their development, and also because of the ability of investors to more effectively intervene and manage a given investment.

Nye used the example of the risk of psychiatry and brain-based medicines made evident in the "abandonment" (Stowall 2011) of the CNS sector by a large number of companies as part of what, in his narrative, was a strategic corporate opportunity. Importantly, he also made explicit what had been driving the need to shift to early-stage investing and partnering: the pharmaceutical industry's abandonment of the commercial neuroscience sector. In this way, we see the relationship between early-stage partnerships and sector divestment.

Outside these insider spaces, few people were aware that something as seemingly ubiquitous as the everyday antidepressant was on the verge of collapse—new investment, that is, was drying up. A dramatic transformation in the way psychiatric and neurological therapeutics had been envisioned was unfolding. Given the growth and evolution of psychobiological theories of mental illness stemming from the cognitive revolution of the 1950s (Gardner 1985; Bateson 2000), the global expansion of psychophar-maceutical models throughout the twentieth century (Lakoff 2005; Petryna and Kleinman 2006), and their authentication within biomedical narra-tives about mental illness (Gordon 1988; Rose 2004), the antidepressant had expanded to become the de facto solution for depression within global biopsychiatry by the end of the twentieth century. By 2011, however, the landscape had changed dramatically with the abandonment of in-house neuroscience research and development by pharmaceutical companies.

During Jeffrey's subsequent presentation at the 2011 Neurotechnology Investing and Partnering Conference in San Francisco, he compared John-son & Johnson's decision to stay with neuroscience to the act of "running into a fire." He said, "First I want you to feel *good* about yourselves. ... I feel that this audience and the [conference] organizers are running into the fire to rescue people with neurological diseases when people [pharmaceutical companies] are abandoning them." By 2011, this abandonment was all too apparent. Six months after the 2011 conference, the Swiss pharmaceutical company Novartis announced it was closing down its neuroscience facility and planning to create a small group in Cambridge, Massachusetts, focused on neurogenetics (Abbott 2011). By December 2011, several big pharmaceu-tical companies—AstraZeneca, Merck, and GlaxoSmithKline—significantly downsized or entirely closed their neuroscience research divisions (Cressey 2011; Smith 2011). When AstraZeneca and GlaxoSmithKline closed their internal neuroscience research infrastructures, they also retreated from their psychiatry research, shuttering any new research programs and prod-ucts for depression and schizophrenia, constituting a "mass exit by many pharmaceutical industries from the sector of the central nervous system disorder (CNS) drug development" (Van Gerven and Cohen 2011, 72). The exit from CNS was also a retreat from multiple neuroscience-related areas of research in psychiatry, neurology, Alzheimer's, and anxiety disorders.

This massive shift also had a global dimension: in concert with global psychopharmaceuticalization, there had been subsequent globalization of these psychiatric models that dovetailed with the creation of psychiatric markets. Given that these markets were shaped so strongly by pharmaceu-tical companies (Applbaum 2009b; Biehl 2007; Healy 2004; Petryna 2009; Petryna and Kleinman 2006) and that many of these companies were in the United States (Nutt and Goodwin 2011), questions remained about the implications for global health in the aftermath of Pharma's abandon-ment: What would be left of the psychiatric models exported globally and then abandoned by their progenitors? What would be the consequences for patients, whose identities and social worlds would now be thoroughly implicated in existing psychiatric discourses?

Well before Novartis's 2011 announcement that it was dropping CNS, the French pharmaceutical company Sanofi-Aventis also dramatically scaled back its neuroscience division, ending the development and research pro-grams for more than a dozen pharmaceuticals in 2009. Ironically, in 2009 Sanofi hired Elias Zerhouni, the former director of the NIH who largely spearheaded the push toward TN at US universities. Not long after, Sanofi partnered with a university to do the work of its now-shuttered internal laboratory.

Were university TN arms morphing into externalized R&D arms for pharmaceutical companies? Soon after AstraZeneca announced in 2012 that it was shuttering a large part of its own internal neuroscience R&D arm, it also announced the creation of a new strategy: rather than a whole-sale retreat from neuroscience, AstraZeneca was, in its words, transforming its approach to innovation. What was the company's solution? AstraZen-eca created an entirely virtual unit called the Neuroscience iMed, part of its IMED Biotech unit. Organized around small virtual teams around the world, AstraZeneca researchers would work directly with academics, rather than solely researching in-house as was their old model. According to Frank D. Yocca, vice president for Strategy and Externalization for AstraZeneca, this infrastructure virtualization was a partnership model:

> As a general rule, projects were always prosecuted internally. True collaborative efforts with academics and biotechs were the exception. In the new Neuroscience iMed, all projects will be engaged through collaborative efforts with academics, biotechs or CROs [contract research organizations]. ... From a Neuroscience iMed perspective, we are operating in a new model where almost all our activities are conducted through CROs and partners. (Vinluan 2012)

The virtualization of AstraZeneca's innovation arm meant that the entire world became a potential collaborating laboratory. This has implications for understanding the future of global innovation and the export of highly specific research agendas around the world. Yet it also reflects transfor-mations in global labor practices: akin to the "thousands of independent software developers" that Apple and Google "leveraged" according to the report from Booz Allen (now part of booz&co. as of this printing) (Le Merle and Campbell 2011, 3), outsourcing innovation and growing labor casualization under neoliberalism means this virtual innovation center is both expanding and globalizing while it exercises the neoliberal management models of downsizing, outsourcing, and constricting. AstraZeneca, like many of the large Pharma companies that closed their internal research programs, fired many of its full-time employees while creating this new virtual model. AstraZeneca's move was a new strategy among biopharmaceutical companies; the company's model, appeared to *entirely* externalize its neuroscience research structure.

Even more surprising, in 2012 the National Center for Advancing Translational Sciences (NCATS)—the NIH agency responsible for the institutional birth of translational research and science in the United States—announced its new pilot program, *Discovering New Therapeutic Uses for Existing Molecules*. In this program, the NIH spent $12.7 million to match academics with pharmaceutical companies. As part of the program, the pharmaceutical companies would lend their unused compounds to university researchers trying to create new therapies. Especially interesting is that the list of pharmaceutical companies involved in the pilot program included the very companies that had shut down or reduced their internal neuroscience R&D arms: AstraZeneca, Bristol-Myers Squibb, Eli Lilly, GlaxoSmithKline, Janssen Research & Development LLC, Pfizer, and Sanofi (National Center for Advancing Translational Sciences 2013).

As evidenced both by the shuttering of internal neuroscience R&D programs and the subsequent partnering and outsourcing of formerly internal commercial activities to the nonprofit university—partnerships facilitated by the NCATS program—university TN programs quite literally appear to have become the biopharmaceutical companies' R&D arms. Yet the story of Pharma's outsourcing was part of a broader sector trend, one that requires delving into a history of pharmaceutical R&D before 2012.

## 3.3   Financing Pharmaceutical Innovation

The relationship between commercial R&D and national innovation in the US has undergone radical transformation since World War II (Mirowski 2011). While pharmaceutical companies had historically invested heavily in their own commercial laboratories for discovery-based research, these companies downsized or closed many of their internal CNS laboratories and downsized or eliminated research positions starting in 2005. This closure is curious in view of the overall growth in global biopharmaceutical R&D spending between 2006 and 2015, from $101 billion to $141 billion spent.[1,2] Given the fact that many biotechnology and pharmaceutical companies increased their R&D investments, why was there such a rush to downsize R&D and even close whole departments, especially in life science fields such as neuroscience? The answer appears to hinge on three factors: (1) growing shareholder expectations for a return from R&D investments (made more urgent, perhaps, as a result of the global financial crisis), (2) the rising costs of R&D as compared to a low number of new molecular entities, and thus (3) the need to create strategies to lower R&D risk and costs.

Jeffrey Nye underscored the sense of opportunity presented when companies partner with university-based laboratories at the research stage—that is, prior to "publication." Jeffrey's testimony about getting involved before "the race when publications start appearing" clearly implicates laboratory research, or even the moments of scientific ideation that precede it, as a prime stage for instigating commercial involvement. An important characteristic of corporate partnerships with university researchers at this stage is the untangling of epistemological knots—the (perhaps *metaepistemic*) space where researchers, using the language of anthropologist Mary Douglas, consider "options for following this line of inquiry rather than that, of referring to these other works or omitting them" and where "the categories of valuable and useful areas of work are identified, ranked and bounded, elements assigned to classes and sub-classes" (Douglas 1973, 120). The relegation of early-stage university research to the realm of corporate commercial discovery produces a specific epistemological question about TN laboratories: How might this focus on early-stage interventions reframe laboratory research and affect laboratory practices?

Given the discovery-oriented ethos at work in the institutionalization of TN, TN laboratories enable optimal legibility between university-based laboratory discoveries and the "discovery engines" created by corporate partners.[3] For example, in 2011, Johns Hopkins's Center for Biomedical Engineering Innovation and Design announced a partnership with Johnson & Johnson Innovation (n.d.) for the creation of a Technology Accelerator Fund. The Johnson & Johnson funds would be used to "support innovative translational research and development activities, with a focus on medical device development" ("CBID Establishes Technology Accelerator Fund" n.d.). Johns Hopkins even matched Johnson & Johnson's $250,000 investment. According to a press release, the funds would focus on "translating early innovative designs into high-quality prototype solutions." In this way, Johnson & Johnson not only created access to a pipeline of (and a competitive advantage regarding) potential innovations; they also externalized the intellectual labor involved therein to the students at Johns Hopkins.

This single example shows how externalization enabled lower opportunity costs associated with the discovery process (since this work did not take up time and energy from Johnson & Johnson's internal researchers), created less intense competition for licensing university-based innovations (or exclusive access to said research), and functioned as part of an externalization system for the firm's internal discovery process, which according to Jeffrey (2007) also included "biotech companies to do the experiments" once promising knowledge came to light. In this way, Johnson & Johnson effectively externalized a large swath of the scientific discovery process from the intellectual labor of the scientist all the way through to the clinical trial managed by a clinical research organization.

However, the Johns Hopkins example also shows how an early-stage intervention allows greater capacity to influence the construction of the very science that may inform the venture resulting from the biomedical finding. The Technology Accelerator Fund, brought to Johns Hopkins's biomedical engineering students as an educational opportunity, also brought translational modes into everyday learning practices and made those objectives a *starting place*. Nascent educational objectives take shape in a similar manner. Research methods, priorities, and educational outcomes for university-based science work now prefigure later development activities such as product development or intellectual property licensing. In this sense, entrepreneurial thinking becomes a precursor to other kinds of thinking for biomedical engineering students participating in the program. One can now see how translational thinking constitutes a set of investigational and eseentially, epistemological practices born from a set of key presumptions.

To think of laboratory work as part of the corporate strategy is also to consider how the aegis of "innovation" enables stakeholders to fold all kinds of agendas into local scientific practices. For example, Jeffrey was part of the Johnson & Johnson subdivision called *Johnson & Johnson Innovation*, which, even in its own self-description, fastened together innovations in finance (such as creative deal structures) as part of its work of supporting innovations in the "healthcare innovation ecosystem":

> With the current pressures on the healthcare innovation ecosystem, it's more critical than ever to find new ways to accelerate scientific breakthroughs to benefit patients worldwide. By working together with scientists and entrepreneurs at universities, academic institutes and start-up biotech companies, and by employing collaborative and flexible deal structures, we seek to identify and invest in a wide variety of prior to clinical proof of concept innovations. We will be a part of the innovation community and share a diverse range of expertise and resources.[4]

Here again, financing innovations such as "flexible deal structures" and early-stage diverse partnerships "with scientists and entrepreneurs at universities, academic institutes and start-up biotech companies" were narrated as *strategies*. This view of externalization as a corporate strategy is important in order to think through the university's role under new global innovation regimes. The reference to benefiting patients worldwide notwithstanding, creating financial innovations such as "flexible deal structures" to "identify and invest in a wide variety of prior to clinical proof of concept innovations" supports the argument that these partnerships are part of an early-stage investment strategy. In this way, externalization brings into view the way that TN, in its operationalization at universities, constitutes a form of finance, especially in the way that such strategies enable shareholder value creation through lowering R&D costs and risks. New partnership modes reflect financial and investment considerations, rather than merely interests in inspiring health innovations.

In light of these financial considerations, it becomes crucial to reconsider the meaning of partnership under translational science and medicine. Across the spectrum of neurotechnology-interested parties found at the Partnering events, the language of sharing and partnering was everywhere, yet each constituent group at the events seemed to define partnership differently. Concepts such as "sharing" and "collaboration" created particular strategic meanings for biopharmaceutical companies. Although innovation narratives tended to underscore notions of democratic progress, unfettered openness, and sharing, especially to provide access to the Global South (Tsing 2005), this ostensibly open, sharing-focused innovation model uses external partners such as a university as an innovation *source* for ideas and products to be commercially exploited, licensed, and patented. This is not to say that the university does not benefit in partnership models with private industry. But the university work is recast as an extension of larger commercial partnership strategies.

However, the surging focus on innovation in neoliberal economies—part of what Kaushik Sunder Rajan (2006, 133) called the "ideology of innovation"—has increased the need to think critically about how networks created to share and collaborate "openly" (Kelty et al. 2008) may also obfuscate the ways they articulate existing power networks, produce new information privatization and control systems, or enable sharing and outsourcing risk. In fact, the insider term *externalization* may be a much more appropriate descriptor for the articulations and restructurings that occur under the aegis of "innovation."

## 3.4   Externalization at the University as a Risk Strategy

What is *externalization*? In 2011, Booz & Company (which was later acquired and renamed Strategy&), a leading global consulting firm, published a white paper, *Building an External Innovation Capability*. In the section "Harnessing the World's Innovators," the authors offer the following description of the benefits of an external innovation system:

> Many companies can now point to external innovations that have dramatically increased their competitive advantage. Apple and Google leverage the power of tens of thousands of independent software developers to build portfolios of hundreds of thousands of applications for their products to showcase and their users to buy. Life sciences leaders such as Amgen and Pfizer are collaborating with external scientists around the world to push forward the boundaries of biopharmaceutical progress. For these companies and others, the benefits are already clear: an enormous universe of innovators working on technologies, products, and services of relevance to the company; reduced costs associated with moving a larger portfolio of innovations down the pipeline; reduced risk as others put their human capital to work on risky propositions; and accelerated time-to-market as innovation is freed from the shackles of the cumbersome large-company financial, planning, and pipeline process. (Le Merle and Campbell 2011, 3)

The externalization of innovation in this context includes the appropriation of diverse and dispersed labor and human capital (e.g., Google developers and Pfizer external scientists) as well as the expansion of the commodification networks through which innovations can be brought into the realm of a company's discovery engine. In that sense, it also constitutes a growing trend toward the outsourcing of research in the life sciences—a solution that biopharmaceutical companies have already enacted for clinical trials via CROs (Petryna 2009). Yet externalization also intersects with and likely benefits from global labor rearrangements such as the growth of independent contractors, use of temporary labor forces, and global outsourcing—dislocations that produce highly precarious workers under neoliberalism (Molé 2012).

The Booz & Company (LeMerle and Campbell 2011) document also reveals the enactment (even if rhetorical) of an anticorporate narrative in which innovation is "freed from the shackles" of big business bureaucracy. Such narratives often blatantly negate the inevitable commercialization and privatization that happen once corporate partners copyright and patent outsourced innovation work. Thus, fragments of a libertarian DNA may undergird many innovation policies and forms of rhetoric (Barbrook and Cameron 1996; Zlolniski 2006); organic innovation "hindered by regulation" could now emerge "naturally" and meritocratically through the ostensibly open and free exchange of ideas.

Juxtaposed with this conflation of externalization and innovation also lies an explicit playbook of sorts for how and why this mode of innovation externalizes the risks and costs inherent in corporate product development. In the same Booz & Company publication, the authors outline how externalization enables "reduced costs associated with moving a larger portfolio of innovations down the pipeline." In this sense, externalization allows firms to partner with universities at various points in the R&D process without having to take on the labor and costs of the entire process. Yet they identify the most relevant component of externalization by stating that externalization enables "reduced risk as others put their human capital to work on risky propositions" (Le Merle and Campbell 2011, 2).

In addition to an overview of externalization's benefits, the white paper provides an external innovation toolkit that executive leaders can use to create or embolden an "external innovation capability." The toolkit (see figure 3.1 for a snapshot) outlines 15 tools, including strategies such as leveraging innovation networks and later-stage tools such as acquisition programs. Several tools involve university-industry partnerships, including grants and scholarships, innovation networks, joint R&D agreements, corporate venture funds, and in-licensing programs. The two tool descriptions below show how the university is implicated:

**The External Innovation Tool Kit**

Though by no means exhaustive, the chief innovation officer's external innovation capability tool kit draws on a variety of tools at each stage of the innovation value chain:

1. **Grants and Scholarships:** Providing research grants to education or government institutions can leverage outside expertise with smaller investments. Funding innovative organizations can also provide a company with new technology and ideas from the diverse set of companies and stakeholders often brought in by these external bodies.

2. **Innovation Networks:** An innovation network can enable a company to develop an external community of interest within a framework controlled by the company that is open to outside innovators but closed to competitors. The goal is to encourage problem solving by a systematically identified network of global thought leaders in areas of interest to the company, as Amgen has done with its "extramural research alliances."

3. **Joint R&D Agreements:** These agreements, typically made with academic institutions or government agencies, are designed to support the mission of each partner. Such agreements are typically structured as long-term partnerships focused on areas of mutual interest to both parties. Potential benefits include breakthrough university research or government insight into industry challenges.

4. **External Advisory Boards:** An external advisory board is a multi-stakeholder group made up of different businesses and industries, often with representation from government or academia as well. The external members of such boards are usually independent experts with a diverse set of specialized expertise or knowledge. It is typically easier for a company to introduce expertise through an advisory board than to hire similarly experienced resources.

5. **Certified Developer Programs:** Certified developer or user networks can create communities of loyal partners, helping to accelerate and improve development and adoption. Aggressive adopters like Apple and Google have pioneered the reliance on third-party ecosystems to develop core products far more quickly than competitors that focus on in-house development.

6. **Incubators:** Incubators typically work with early-stage startups by providing facilities, contacts, and business advice to help them prove and launch their business. Partnering with an incubator can provide valuable access to potential technology partners as the startups develop their products.

7. **Corporate Venture Funds:** Similar in operation to a traditional venture capital fund, this type of fund is run by an individual company and uses the company's capital to invest in promising technology. It allows the company to exercise more control and to participate directly in the success of the companies it invests in. Corporate funds typically invest in companies with close tie-ins to the investing company's strategic goals, rather than for purely financial reasons.

8. **Limited Partnership Positions:** Investing in a traditional VC firm as a limited partner allows a company to make focused investments in particular technologies while spreading some of the risk with the general partner and other limited partners. This technique also takes advantage of the general partner's access to deal flow and expertise in negotiating and structuring deals and crafting exit strategies.

9. **Related Ecosystems:** Related ecosystems usually come in the form of alliances or joint ventures among companies joining together to support a technology or to package products into a solution. Cisco, EMC, VMware, and Intel have recently created such an ecosystem by contributing some of their intellectual property and product expertise to launch VCE, which targets the virtual computing environment. These ecosystems allow individual companies to leverage partner companies' capabilities and potentially gain access to untapped capabilities or resources, along with new customers in adjacent markets.

10. **Collaborative Innovation:** These efforts come in many forms, ranging from collaborations with individuals to working with other leading enterprises. Companies can use external innovation to collect ideas from a variety of sources, which might include external communities of enthusiasts, such as the open competition sponsored by Netflix to discover improved algorithms for recommending movies; vast networks of internal employees like Procter & Gamble's; or other companies in related fields.

11. **Venture Exchanges:** Venture exchanges provide a structured environment in which participating venture funds can feature their portfolio companies. Typically, a group of venture funds agrees to participate on a regular basis, although some venture exchanges are ad hoc in nature. Funds showcase companies with the requisite focus to potential customers, partners, acquirers, the press, and others, depending on the venture exchange's focus.

12. **Key Opinion Leader Councils:** These councils typically resemble an external advisory board in makeup, but they differ in scope, focusing on the problems and needs of the broader industry rather than on the direction of a single company.

13. **Joint Ventures and Partnerships:** Joint ventures are formal arrangements in which two companies form a legal structure and combine a defined set of assets and capabilities to jointly address a specific market need. This arrangement is particularly useful when the companies have complementary capabilities but neither could go to market on its own. JVs are one step removed from a merger, in that the parties are still independent but have very closely allied themselves in the chosen market.

14. **In-Licensing Programs:** In many cases, companies find it more cost-effective to license intellectual property that has already been developed than to develop their own. A company typically makes such a deal to enhance the functioning of its current products, or to use its superior commercialization capability to bring new technology to market more effectively than the technology owner could.

15. **Acquisition Programs:** When a technology that rounds out a company's product portfolio or strategic position is developed by another company, one avenue is to acquire sole ownership in the technology and/or the company that has developed it. Working with corporate development teams, CIOs can put in place technology acquisition programs that will scan the external environment for promising targets and bring them in-house.

Each of these tools can be quite effective when properly matched to the user's technology and market conditions. The key is to employ the right set of tools to maximize the probability of success without spreading management resources and attention too thin.

**Figure 3.1**

A section of Booz & Company's external innovation toolkit (Le Merle and Campbell 2011, 4)

*Grants and Scholarships*: Providing research grants to education or government institutions can leverage outside expertise with smaller investments. Funding innovative organizations can also provide a company with new technology and ideas from the diverse set of companies and stakeholders often brought in by these external bodies.

*Joint R&D Agreements*: These agreements, typically made with academic institutions or government agencies, are designed to support the mission of each partner. Such agreements are typically structured as long-term partnerships focused on areas of mutual interest to both parties. Potential benefits include breakthrough university research or government insight into industry challenges. (Le Merle and Campbell 2011, 4)

As shown in figure 3.2—the model provided by Booz & Company—the highest level of risk in the product development life cycle lies precisely at the point where the tools employed involve partnerships with universities. One sees here how early-stage externalization models enable private firms to "outsource" some of the *riskiest parts* of the R&D process to universities. Thus, in a context in which "externalization" is an explicit corporate strategy, TSM looks to be a thinly veiled externalization program in which these externalization-focused industry-university partnerships in fact function as a system by which high-risk investments are de-risked prior to potential

Figure 3.2
Booz & Company's external innovation tools analysis. The upper left shows how the riskiest part of the product development process corresponds to "tools" closely associated with university partnerships. Although the profile refers to the risk of failure, multiple risks make early-stage innovation even more challenging (Le Merle and Campbell 2011, 7).

later commercial investment. At the same time, the externalization itself constitutes a significant and self-contained financialized action whereby Pharma's R&D costs and risks are lowered, access to potential discoveries maintained, and short-term shareholder value delivered. The externalization of science and medicine was already a financial outcome—without any consideration of actual (e.g., patient) outcomes that come or might emerge from translational partnerships.

## 3.5  Becoming Translational: The University as a Risk Shelter

In a social-level analysis, there is a way to think about the transfer of risk achieved through TSM as also being a transfer of risk back to the public via the nonprofit university. The emergence of the university as a space of *innovation in science and medicine* cannot be taken as a mere reflection, as several scholars have suggested, of the animations of capital and capitalism alone. Rather, as is detailed here, a very important context informs the sudden demand that universities become translational. As an entity of the state, the NIH used public research funds to engineer a translational infrastructure to be housed at nonprofit universities, which, because of their tax structure, are also effectively semi-publicly supported. What emerges is the sense of the public sphere as a space of private sector discovery and innovation and, as I argue in what follows, an opportunity to offload private sector risk.

As part of this logic, it is important to understand how the university is envisioned as the space for this kind of innovation. Quite a few commenters support a stronger role for the university in pharmaceutical research and development, including many academics critical of biopharmaceutical companies' knowledge practices (Applbaum 2009c; FitzGerald 2005; Gomez 2006; Insel 2009; Minogue and Wolinsky 2010). Often, these arguments position universities as a necessary reprieve from the for-profit ethos responsible for the CNS innovation crisis. The proposal by many of these commenters to separate drug development into "humanitarian" research and scientific drug development areas reflects this ethos and outlines appropriate roles for universities and pharmaceutical companies.

Stakeholders have seen the university as perfectly positioned to contribute to the ecosystem of biomedical innovation because of the university's focus on "good science"—the issue frequently used in explaining the failure rate in bringing neuropharmaceuticals to market. Executive

statements about "good science" seemed to imply an implicit acknowledgment that low-quality science went into the in-house development
of new psychopharmaceuticals. During a cocktail hour at the 2011 Neurotechnology Investing and Partnering Conference, one pharmaceutical
executive confided to me (after several drinks) that CNS (the neuroscience sector) had to change; larger shifts in the pharmaceutical industry
coincided with opportunities offered by the university partnership model.
He stated,

> We can't do it like we have been doing in the past—finding a drug and just mar-
> keting the hell out of it. We're trying to understand the diseases in a much more
> biologically rigorous way, and we're looking for a signal, and then we'll look
> forward. This is why academic partners are useful. They are patient and take a lot
> more time than we take in the private sector. (Interview 24, May 21, 2011)

It was in the pharmaceutical companies' interest for universities to assume
responsibility for basic research into the neurobiological mechanisms in
Alzheimer's, for example, since this lack of knowledge (perhaps combined
with overpromising neuroscience's new insights) appeared to be a sustained
challenge for pharmaceutical companies (Belluck 2011; Graham 2006;
Lock 2013) in commercializing and developing new compounds and drugs
(Minogue and Wolinksky 2010).

However, the partnership between universities and industry is not one-
sided. University-based translational science centers have the resources to
meet industry's needs, but they also benefit from partnerships with industry, especially after enduring large funding cuts in the late twentieth and
early twenty-first centuries (e.g., billions of dollars cut by the 1997 US Bal-
anced Budget Act), decreased clinical-service reimbursements for academic
medical centers (resulting from a rise in uninsured patients using academic hospitals), and shifting federal funding (Karlin 2013). Universities
are under increased pressure to transform their mission toward developing
applied learning, marketable skills, patent creation (Mirowski 2011), and
corporate partnerships. The modern nonprofit university is under explicitly
neoliberal pressures (Bok 2003; Raaper and Raaper & Olssen 2015; Strathern 2000) despite its long history of instrumentalist concerns (Brett 1945).
Additionally, that universities have historically relied on a stable demarcation from for-profit corporations challenges the notion that they ought to
become the space to reclaim big science. The growing evolution of the university and the increasing trend toward partnerships with pharmaceutical

companies—clearly evidenced through the recent NIH-mediated program matching unused compounds with academic partners—suggest the non-profit university is emerging into a hybrid institution.

Moreover, in the sense that pharmaceutical companies "outsource" to public universities via public institutions such as the NIH, the public has already become a site and means for the remediation of private-industry concerns. The evolution of public-private partnerships and assimilation of corporate models into public institution infrastructures may suggest that "the public" may enable a hiding space for the interests and risks of the private. Assumptive ethics about the superiority of a more democratic pub-lic science (in which research and discovery would be the domain of the university instead of pharmaceutical companies) dismiss the reality of new, complex formulations that inhabit and potentially remake the entity that is the public institution.

I argue, therefore, that given the acknowledged risks of early-stage TN research and the intentionality with which for-profit companies discharge the riskiest parts of their research process, this shift turns universities into risk shelters. In a *New York Times* op-ed, a Cornell University psychiatry professor extolled the benefits of bringing pharmacological research to the university and the university's suitability for this kind of research: "The pharmaceutical industry has little taste for the long-term financial risk of discovering new psychotropic drugs. ... In contrast, academic researchers, who are not beholden to shareholders, are much freer to do high-risk, high-gain experiments that can fail. ... These discoveries, in turn, may entice the drug makers to reinvest in psychiatric drug development" (Friedman 2013).

The author, while correct about the distaste for risk by pharmaceuti-cal companies, is wrong on several other fronts. He offers an unrealistic, idealized view of the university—one that neglects the reality of increased rationalization and demands of bureaucratization (Shore and Wright 1999). Under neoliberalism, the university is no longer "nonprofit" in exactly this way. The author also problematically assumes that pharmaceutical com-panies have cut all ties to the development pipeline. This is not the case. Pharmaceutical companies are very much involved in retaining licensing options and the emergence of industry-university partnerships; programs such as Pfizer's Centers for Therapeutic Innovation (CTI) and AstraZeneca's iMed unit illustrate this emerging reality. In fact, the translational shift at the university may very well bring "shareholders" into the fold of academic science, effectively broadening the structure and ecosystem of translational science and medicine. But in the sense Friedman articulates, the university could sustain risky investments and failures in science and experiments in ways that a for-profit company could not. However, the notion that universities can fold risky pharmacological research into their larger mission, and that university research without a yield is *merely education* under the university's educational mission (rather than a *loss* for the pharmaceutical company), reflects an important, if misguided, amalgamation: the educational mission of the university and thus its nonprofit status justifies the import of for-profit research modes from newly risk-wary biopharmaceutical companies. Here conflations occur: the articulation of the university as risk immune (i.e., nonprofit) while possessing indefatigable risk capacity, with the university as both outside the market and yet also perfectly conducive to market processes. The assumption about risk presumes that the risks taken on by universities simply disappear therein rather than recognizing the complex financial costs incumbent with the new partnering.

A more significant issue is the way that risks emerge in the translational paradigm. As universities take on the risk-laden wing of translational research and development and corporations commercialize university products, the university's role transforms in relation to private investments in risky bioscience. A primary contention here is that in this risk exchange at work in TN's public-private partnerships, the university becomes a *de-risking mechanism* for high-risk neurotechnology investments. That is, the university ostensibly improves or cleans high-risk, often early-stage investments (investments made after the scientists or biotechnology company ostensibly have proved and de-risked pharmaceutical potentials for development through strong empirical research). Thus, if accomplished via external partners such as the nonprofit university, the process essentially cleanses a pipeline of discoveries for financial investment. In this sense, federally funded translational science and medicine and the restructuring of academic science become part of a growing *financial mechanism* in which high-risk investments are de-risked via the public sphere.

This is not to suggest that universities are merely exploited.[5] Indeed, universities potentially benefit from licensing "effective" discoveries to pharmaceutical companies (FitzGerald 2005), and provisions of legislation such as the Bayh-Dole Act mandate that for-profit organizations that license university technologies pay market value. However, the on-the-ground reality is more complicated. For example, Northwestern University made a total of $1.48 billion in revenue from licensing agreements between 2002 and 2011 (Wang 2012). A large amount of the revenue was due to the success of the drug Lyrica, the blockbuster antiseizure pharmaceutical developed from the work of a Northwestern University chemistry professor during the 1980s. A local Chicago newspaper (Wang 2012) chronicled the story:

> Lyrica was better than a good idea. In the 1980s, Northwestern chemist Richard Silverman developed a molecule that limited epileptic attacks. The drug produced $3.7 billion in 2011 sales for New York–based pharmaceutical firm Pfizer Inc., which started marketing Lyrica in 2005. Northwestern received a onetime, $700 million payment as well as big annual fees from the drug's sales. The university's gross licensing revenue reached $191.5 million in 2011, "virtually all" of it from Lyrica, officials say.

Despite these staggering revenue figures, the norms for licensing revenue were much less profitable, and blockbusters for universities and pharmaceutical companies alike were rare (see figure 3.3). The same newspaper article quotes Eugene Sunshine, then Northwestern's senior vice president for business and finance, as he explained the rarity with which scientific

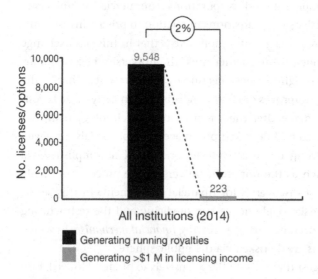

**Figure 3.3**
The number of licenses/options generating running royalties compared to the number of royalties or options that produce income in excess of $1 million annually.
*Source*: 2014 AUTM Licensing Survey.

research could be translated into significant profits: "The track record is not great. ..." While 2016 saw an increase in revenue, only 2% of running royalties produced more than $1 million per year (Reslinski and Wu 2016). The relatively low revenue from university technology licensing (and the low number of applications that emerge from academic laboratories in general) also create a problem for the subset of narratives that justify public-private partnerships based on some simplistic linear notion of economic productivity.[6] Thus, an important question emerges: Where there is typically little revenue and low financial yields from such partnerships, what are the consequences for a university newly burdened with costs and/or risks that were outsourced through industry collaborations?

In an R&D timeline replete with risk and with the incredible difficulty of mapping the brain, existing CNS failures have been categorized as mere scientific problems (Fang and Casadevall 2010), and the university is envisioned as the solution by defining CNS failures as knowledge problems. However, imagining the university as a solution still relies on the view that psychopharmaceuticalization failures and possibilities for improved health are principally about science. In other words, a scientific determinist view would see "real" (rather than corporate) science as the solution to unearthing the biological bases for mental disorders. In this sense, externalization to the university would be a scientific homecoming as well as a preservation of certain models of mental order and disorder. However, this is not the case.

## 3.6  There Are No Biomarkers: Understanding the Retreat from Psychopharmaceuticals

I argued that TN's systemic function at the social level is as an investment de-risking tool. As TN turns the research university into an innovation engine and the infrastructure of TSM into a de-risking system, it is clear that TN serves a particular financial function for risk-burdened biopharmaceutical companies. The R&D risk involved in neuroscience had rendered it such a significant investment risk that abandonment of the sector by many companies had been inevitable. This process started in 2006. Simultaneously (2006–2010), biopharmaceutical companies exited commercial neuroscience sectors, downsized or closed down their internal laboratories, and TN centers emerged at universities. The institutionalization of TN at the university thus bears examination against an emerging context of risk that is historically situated and inextricable from a set of specific healthcare corporatization events between 2006 and 2010.

It is worth highlighting the context in which major pharmaceutical companies abandoned CNS and the function of externalization programs in relation to this shift. This history is tied to larger knowledge problems specific to psychopharmaceuticals that became known between 2006 and 2010. Getting to the unexpected relationship between "knowledge risk" and financial risk in the CNS sector involves traversing what caused the massive divestment in commercial neuroscience in the first place, including the pharmaceutical industry's exit from the antidepressant field.

One primary concern for commercial neuroscience begins with a simple, intractable problem: the brain. For many observers, the problems attached to brain-based research, especially on brain disorders, have been profound: brain-based clinical research has a high rate of failure across the discovery spectrum and failures have been unusually expensive (Nutt and Goodwin 2011; Cutler and Riordan 2011). Scholars have suggested that finding molecular targets' mechanisms of action can be exceptionally difficult, especially for psychiatric disorders (Insel 2009; Miller 2010). Compared to other pharmaceutical and biomedical therapies like cardiovascular interventions, the failure rate for therapies targeting CNS disorders is particularly high (Cutler and Riordan 2011). The authors of a report from the 2011 European College of Neuropsychopharmacology meeting cataloged problems that plagued R&D pipelines as follows:

> With an average of 13 years, the time to develop a medicine for a psychiatric indication is longer than that for other disease areas. ... The failure rate of medicines in psychiatry and neurology is higher than that for other disease areas and many medicines fail late in the development process—at Phase 3 or even at registration—leading to particularly high financial loss. (Nutt and Goodwin 2011, 496)

Beyond the difficulty of research and development lies the challenge of trying to create models for psychiatric and neurological diseases. In other words, problems arise in trying to understand the underlying neurobiology of disease states and turning neurobiological knowledge into safe, effective therapeutics. Although clinical neuroscience research tends to narrate this difficulty using the language of biological complexity, the unique and multimodal nature of *mental illness* also renders problematic the very categories and methods used in bioscience research. According to the report, challenges in understanding the biology of psychiatric disorders have hindered additional research using existing biological conceptualizations and methods:

> Predictive and prognostic biomarkers for psychiatric disorders are largely nonexistent, and the poor predictability of the pre-clinical models for both psychiatry and neurological diseases is only now being addressed by the development of experimental human models. Most psychiatric medicines have been discovered by guided or "well-educated" serendipity. (Nutt and Goodwin 2011, 496)

Accordingly, for many observers it is the combination of early-stage roadblocks (Cutler and Riordan 2011), especially concerning poor or insufficient scientific understanding (Cressey 2011; Wheling 2015), alongside challenges in development, clinical-trial problems, and regulatory hurdles—many of which themselves are considered consequences of scientific challenges—that make CNS research so difficult (Garcia-Rill 2012). Because "only around 8% of CNS drugs that make it to clinical trials end up being approved" (Miller 2010 503), the 92% failure rate for the larger enterprise shows just how costly CNS is. From an investor's perspective, the enterprise of CNS research is overloaded with risks—particularly knowledge risks.

Although there has been broad consensus among neuroscientists, investors, translational researchers, academic medical center executives, pharmaceutical executives, and patients about the many complications along the pipeline between the laboratory bench and the patient bedside, pointing to the existing problems and challenges with CNS research fails to address the inefficacy of many of the psychopharmaceuticals marketed in recent decades. In a provocative article, the director of the National Institute of Mental Health (NIMH), Thomas Insel (2009), compiled research to claim—controversially—that new antipsychotic psychopharmaceuticals are not significantly more effective than older ones. In the case of antidepressants, there is little evidence, according to Insel, that existing first- or second-generation pharmaceuticals are effective at all compared to placebos:

> Studies have begun to examine the value of newer, expensive medications relative to older, less expensive, and less commonly used compounds. For the antipsychotics, there are now four large-scale studies demonstrating the lack of superiority of the second-generation compounds over first-generation compounds. ... For the antidepressants, with either first-generation or second-generation

medications, the rate of response continues to be slow and low. In the largest effectiveness study to date, with more than 4,000 patients with major depressive disorder in primary care and community settings, only 31% were in remission after 14 weeks of optimal treatment with the second-generation medication Citalopram, a selective serotonin reuptake inhibitor. ... In most double-blind trials of antidepressants, the placebo response rate hovers around 30%. (Insel 2009, 702)

Insel's analysis suggests that the new class of medications that Big Pharma brought to market in the late twentieth century perform no better than off-patent pharmaceuticals, which indicates a failure of pharmaceutical or pharmacological innovation. Moreover, the analysis suggests poor efficacy among *all* antidepressants. In another controversial meta-analysis published in the *Journal of the American Medical Association*, Fournier and colleagues (2010) suggest that for the mainstream population of users, antidepressants statistically just do not work. Of course, there is considerable disagreement over whether this so-called crisis is an actual one (see Greenslit and Katchuk 2012, and more controversially, Light and Lexchin 2012) and there is even more disagreement over the claims of Insel and others. Nevertheless, there is a clear perception that pharmaceutical innovation problems reflect problems in the biological understanding of mental disorders and attendant psychopharmacology challenges. According to Insel (2010, 503), "There are very few molecular entities, very few novel ideas, and almost nothing that gives any hope for a transformation in the treatment of mental illness." Similar critiques and analyses have been made in other areas, including, again, treatments for cancer (Davis 2017).

Insel's statement historicizes several interesting points. First, an important dualism: the *realization*[7] of the failure of the antidepressant is the articulation of the massive scientific ambiguities on which new psychopharmaceutical solutions were based as well as social and cultural meaning making around the notion of a "successful" pharmaceutical (Greenslit and Katchut 2012). Second, this realization produced a question about the disconnect between the psychopharmaceuticals and their promises. Why didn't the underlying scientific ambiguity slow the growth of psychopharmaceutical markets or advances in existing pharmacological models and neurobiological presuppositions during the late twentieth century?

### 3.7   November 2010: Opportunity via Scientific Uncertainty

In November 2010, I attended a seminar given by Tom Insel titled "Ensuring Public Trust: Conflicts of Interest in Psychiatry." This neuroethics-focused event (see figures 3.4 and 3.5) was part of a panel on global mental health that took place in San Diego, adjacent to the 2010 Society for Neuroscience Conference. I attended this event to learn more about whether, and if so how, pharmaceutical practices were being approached and understood within neuroethics. The primary topics encompassed classic and philosophical concerns such as issues of enhancement, consent, autonomy, and free will. However, there was little discussion of the transformations at work in the pharmaceuticalization of mental health or deployment of problematic

**Figure 3.4**
A presenter at the neuroethics-focused event in San Diego, California, in November 2010

**Figure 3.5**
Presentations at the San Diego neuroethics event in 2010

neurobiological models, and surprisingly little research or discussion about the politics of pharmaceutical practice.

I was especially interested in this event because Insel had published an article a year earlier in which he was critical of the trajectory of TN (Insel 2009). Although much of his writing focused on practical changes needed within neuroscience, he rarely explicitly articulated the political and ethical issues attached to the modern landscape of neuropsychiatry, pharmaceuticalization, and global health. Yet here, Insel would speak about the specific neuroethical issue of conflicts of interest in the relationships between pharmaceutical companies and university-based physician-researchers and about the practices of pharmaceutical companies generally. "I wanted to give a topic that hasn't been brought up," Insel said during the keynote address. He added that even though classic conflict-of-interest issues had not been dealt with in neuroethics literature, "there was a research opportunity here." He went on to describe explicitly what he saw as a conflict of interest in prescribing psychiatric drugs:

> There is a correlation between those [psychiatrists] who receive the funds with the level of prescribing. A fact that's not well-known to the public is this: the number one class of prescribed drugs is antipsychotics. Not cancer drugs, not heart or blood pressure medications, but antipsychotics. Atypical antipsychotics are at numbers 5 and 6, with antidepressants a bit further down. (Informal comment at "Ensuring Public Trust: Conflicts of Interest in Psychiatry," San Diego, November 2010)

The ethical issues according to Insel were locatable in pharmaceutical companies' widespread practice of creating "me-too drugs," referring to the practice by pharmaceutical companies of creating similar or minimally different new pharmaceuticals to profit from patents accorded to that "new" drug. As this practice causes high healthcare costs and is seen as a strategy motivated by profit seeking over actual improvement in patient health, this corporate practice, which continues today, raises an ethical issue. Insel also referenced research that showed a drug's patent status—rather than efficacy or patient benefit—is most highly predictive of prescribing practices. According to him, "Prescribing practices are the primary reason for high healthcare costs in this country and for Medicare and Medicaid—this is part of the problem." Several panelists at the table looked down during Insel's speech. One panelist seemed to stare intently at the intricately patterned rug that adorned the floor in the exhibition space. "There is a tendency to demonize Big Pharma or psychiatry, and that is not my goal," Insel offered. He continued, "There are no biomarkers for psychiatric disorders, which creates opportunities for bad behaviors such as the creation of new diagnoses."

Insel's contention about the social construction of disorders was a rare instance in which scientific ambiguity (here exemplified by the absence of biomarkers for psychiatric disorders) was explicitly tied to exploitative pharmaceutical practices. Of course, this brings up a question about the materialities of these drugs: If they do not directly implement precise knowledge about biomarkers, what presumptions and models might they contain? Might this history also reflect the literal enrollment or embedding of scientific ambiguity into the materiality of emerging neurotechnologies? One can begin to think of the risks of scientific imprecision as a material property of these new drugs.

In his optimism about the presumptions of the problem-free nature of biomarkers, Insel still situated *good science* as a recourse—a protection mechanism against exploitative pharmaceutical practices. This explanatory trend was prevalent among the few neuroscientists and physicians who waded into neuroethics. Informed by a high-minded scientific determinism, they often proposed designs for ethics in which the truth of science was the source and the solution for neuroethical problems. Nevertheless, Insel then went on to provide an instructive insider history of the role of psychiatry as a professional discipline in prescription practices and the implication of academic medicine. According to him:

> Psychiatrists have long been at the low end of the economic ladder. There isn't much that a psychiatrist can make money from. Big Pharma was having a problem of a failure of new ideas and had to find a way to market "me-too" compounds. That was a great business strategy, but it means Big Pharma gets focused on remarketing compounds and the creation of new intellectual property. Pharma has found that appealing to academic prestige and currency leads academics to become marketing agents for pharmaceutical companies, who then have a real impact on physicians. But NIMH [the National Institute of Mental Health] also cares about what clinicians do. We're not funded to help academics write papers. Scientific messages become counteracted, and so the Pharma strategy is good business practice, and their incentive is to keep a patent alive. Most new drugs are "not so new." This happens to be correlated to the rash of new stories about new and old technologies and the problems of efficacy and cost.

Insel then posed the following question: "So, [with a market share of] 25% of the entire pharmaceutical market, antipsychotics are huge sellers for the pharmaceutical companies. Is this a problem?" He paused and searched for a euphemism. He continued, "Might it [at the very least] enable destigmatization?"[8]

The problem, according to Insel, was that off-patent products were equally effective as the new (patent-worthy) ones, so the costs were unnecessary: "The cost of healthcare in this country means that this is an ethical issue." Of course, this particular problem of efficacy for *new pharmaceuticals* is separate from the problem of general efficacy for all psychopharmaceuticals—an issue that Insel outlined in his 2009 article. Interestingly, healthcare costs—rather than development and marketing biotechnologies such as pharmaceuticals based on scientific ambiguity—became the means for Insel to envision this as an ethical issue, leaving aside the question of the general efficacy of psychopharmaceuticals for general populations.

Insel addressed the question written in bold letters across his PowerPoint slide: "Does Promotion *Affect* Prescribing?" According to him, the answer is yes. Yet doctors "simply can't see it." Free samples are about creating a relationship, but physicians are not aware of labeled drug indications. Solution? "Pharma has really stepped up" and disclosed payments to doctors, said Insel, but he suggested that this was not entirely voluntary. Pharma was leaving CNS and trial data to work in other areas. For Insel, this retreat from CNS was anything but amicable. Instead, he saw academics as part of the problem because medical schools had yet to go beyond conflict-of-interest disclosure in talking about physician-researcher embeddedness in pharmaceutical practice.

According to Insel, "This is about something more complex, which is, how do you take the need we have for the next generation of therapeutics [and develop it such] that it doesn't become a marketing exercise in which all of this gets perverted for the gain of a few academics that get involved and ultimately to the violation of public trust, which we're most afraid of losing?" To explain the way that pharmaceutical practices affect clinical outcomes, Insel brought up the example of psychiatrists who expressed interest in cognitive behavioral therapy despite the rising prominence and dominance of psychopharmaceutical treatment modalities, and who, because of their "rogue" partialities, became a source of pharmaceutical disinterest. The problem with this pharmaceutical distancing is that cognitive behavioral therapy was still shown to be *highly effective*, Insel stated emphatically.

His caution drew attention to the ways that pharmaceutical interests produce narrow psychopharmaceutical presumptions, which in turn affect the way one even imagines innovation in research. His caution also pointed to the impact of psychopharmaceutical presumptions on psychiatry, made evident in the denigration of cognitive behavioral therapy. Most importantly, such a presumption likely also shaped the research programs and research interests advanced by the National Institute of Mental Health under the translational shift. Insel's caution also underscored the issue of foreclosing of other potential solutions when thinking about TN's epistemological contours (toward pharmaceuticalization).

In the context in which novelty, intellectual property, and patents are primary goals in neurotechnology innovation, TN may lead to the marginalization of effective, nontechnological, and even effective, extant solutions. During the question-and-answer session (see figure 3.6), one slightly disheveled man asked a rambling question in which he stated, somewhat antagonistically, that he *did not believe* that NIMH was an advocate for cognitive behavioral therapy—in contrast to Insel's exhortation—and that there had indeed been a cultural aspect to this shift. The audience member said that he was a physician, but one could sense the audience's discomfort with this man, whose appearance suggested he was an outsider to these conversations.

According to the man, the NIMH itself had created a culture in which prescribed drugs were viewed as the primary way to treat psychiatric disorders. In response, Insel offered a critical insight:

> When we've talked about these issues, we see that the culture of drugs-as-treatment was created by us, and secondly, when there have been alternatives,

**Figure 3.6**
The question-and-answer session at the San Diego neuroethics event in 2010

it's hard to make a living arguing that CBT [cognitive behavioral therapy] does impact the brain. This model makes academe the primary force and companies as a capitalizing figure, and there is something interesting about this chicken-egg formulation.[9]

Here Insel offered a rather direct admission about the commercial considerations at work. The idea that it is "hard to make a living arguing that CBT does impact the brain" references the clinical encounter for low-paid psychiatrists as well as the difficulties of garnering federal funding to answer certain kinds of research questions about the brain and mental illness. In this sense, Insel was already referencing the institutionalization of a set of epistemologies that have a life beyond concern for patient outcomes alone. Here TN has become a kind of pathway that tethers visions of health innovations to particular kinds of approaches and models. Admitting that certain arguments might make it "hard to make a living" acknowledges how certain psychopharmaceutical modes have apprehended psychiatric, institutional, and scientific thinking. However, the pharmacological presumption is also *technological*. The norm of "drug-as-treatment" thoroughly apprehends how one envisions solutions. It is here that one sees the ethical axis involved in technological determinism. In something like a psychopharmaceutical determinism, the ethics lie not only in the export of pharmaceutical and corporate pressures; the moral issue emerges in defining and articulating available and optimal solutions, and in the avenues available for patients to find official recognition (via diagnosis) of mental health needs.

After the event, I caught up with the man who had posed the question. A quixotic individual, Dr. Stein lived in Santa Cruz, a small left-leaning city on the central coast of California about an hour from Silicon Valley. I asked him about his comments during the event—in particular, what he meant about the culture that NIMH had created. We sat down in an area at the rear of the hall where the session was held. A flurry of attendees was still milling around; a soft murmur surrounded the ten or twelve neuroethics posters set up at the back of the room. Stein explained his experience as a psychiatrist:

I've been in psychiatry for a long time. I used to think that I wanted to be a pediatric psychiatrist only, but then I realized that I wanted to help everyone. My mother had bipolar disorder. Back then, when I was in medical school, they called her a Manic Depressive. I've seen things change and go up and down since the early 80s when I first started out as a psychiatrist. In particular, the part that has changed the most is the tendency to create a diagnosis around everything and, especially working with patients, you quickly learn that the issues that people present can't be neatly fixed with some magic bullet. I really did not like—and a lot of my colleagues at the time resisted—this rush toward the single bullet solution. I mean, for me, I thought, "I really know these people." I know that the issue that they are dealing with is a lot more complex. Now, for the really severe patients, the medications have gotten better and better. For schizophrenia, medication can mean the difference between a life and no life at all. But for the rest? They just don't get a huge benefit from every new drug that comes out and it's been this way for a long time. (Interview 21, November 12, 2010)

"You know, therapy *still* works. I'm one of the few psychiatrists that still talks to my damn patients!" he said with a chuckle. Stein's face was lined. Close up, he looked somewhat frail with unruly hair, tanned skin, and tattered pants. His speech was informal and easygoing, yet also cerebral. He continued:

It's only from talking to my patients that I started to question this whole drive towards some magic molecular bullet. I mean these people have never talked to a person. I can tell you that you [can't] wipe away/explain complex behavior, pain, and abuse with a simple label and a pill. And look, I'm no scientific denialist, I believe in science, and I believe in the brain. I believe that children's brains are impacted if they aren't interacted with as children. I believe in development. I believe in facts. However, I know from years and years of doing this that some simple molecular bullet notion is just ... just the wrong direction. I've had patients that were addicted to all kinds of stuff. Putting them on psych meds just did, well nothing. Nothing! This sales exec came marching into my office some years ago trying to push antidepressants for overweight children. I don't even prescribe it anymore to treat addiction. I think we're missing a whole lot of other options out there because we're so obsessed with this stuff. Sometimes I wonder what we could have come up with. So, nowadays, since I have less clients now, I go and I try to attend these events and I try to get people to think about this stuff. I want to try to change the conversation. (Interview 21, November 12, 2010)

During a different conversation, the well-respected neurologist Dr. Jones had made a statement about the relative standing of psychiatry and neurology. For her, neurology was much more scientific and thus more rigorous. I wondered if Stein's eager contention that he was no "denialist" was about this ongoing assumption/stigma attached to psychiatry as being less scientifically rigorous and if neurology's scientism rendered it more conducive to a sort of biopharmaceutical "embrace" (Good 2007). Indeed, medical anthropologists, sociologists, and historians have long mapped the technical (Adams 2016; Biehl 2013), cultural (Gordon 1988), clinical (Jain 2013), economic (Tyfield 2011, and technological (Lock and Nguyen 2010) means through which health and illness become envisaged as purely or mostly the domains of biotechnology and biomedicine (Clarke et al. 2003; Warner 2014). Nevertheless, the current antidepressant-prescribing practice and its reflection of the role of pharmaceutical companies in clinical practice is quite relevant to questions framed in the laboratory. In this way, Stein's question about potential alternatives is quite compelling.

### 3.8   The Science inside the Psychopharmaceutical

In light of the outcome statistics for pharmaceutical solutions to depression, we might consider the question: What do psychopharmaceuticals contain? There is an opportunity to reenvision the symbolic life of biotechnologies, given that they reflect an amalgamation of scientific practices such as complex theorizations about the brain and *known* knowledge ambiguities, as well as (from a different view) financial practices—including the interplay of trends in global finance and the movement away from high-risk investments during the 2006–2009 financial crisis (Appadurai 1988; Baudrillard 2005). Thinking of these pills as containing these histories opens up research possibilities.

Beyond mere functional analyses of therapeutics or considerations of whether they work or do not work, there is an important consideration for how psychopharmaceuticals have become objects that represent particular forms of modern rationalization, objects that, via the social theorist, Jean Baudrillard (2005, 53), "synthesize gestures of the human in their technical evolution." In particular, there is real relevance in Baudrillard's notion of *functionability*—a concept that does not refer to an object's ability to function per se, but to an object's *manipulability*, which is its capacity to incorporate other social and material forms. Insel's admission that the absence of a biomarker for psychiatric illnesses made psychiatry conducive to problematic behaviors on the part of pharmaceutical companies suggests that psychopharmaceuticals are particularly porous, functionable objects.

Given the relationship between psychopharmaceuticals and the problematic knowledge on which they rest, they have become sites of epistemic risk. However, psychopharmaceuticals are also products of knowledge-related risk taking. Nevertheless, in "A Dry Pipeline for Psychiatric Drugs," the author—a Cornell University psychiatry professor—suggests that the reason for the lack of new pharmaceutical solutions is not merely a dearth of scientific knowledge, but also connects to the complications of a particularly complex object:

> The simple answer is that we don't yet understand the fundamental cause of most psychiatric disorders, in part because the brain is uniquely difficult to study; you can't just biopsy the brain and analyze it. That is why scientists have had great trouble identifying new targets for psychiatric drugs. Also, knowing how a drug works in the brain doesn't necessarily reveal the cause of the illness. For example, just because an S.S.R.I. antidepressant increases serotonin in the brain and improves mood, that does not mean that serotonin deficiency is the cause of the disease; many depressed patients get better with medications that have no effect on serotonin. (Friedman 2013)

Certainly, narratives focused on scientific explanations about the complexity of mental illness refer to its esoteric and biological complexity. Nevertheless, the complexity of mental disorders—and its undoing of various biomedical and cultural framings—often escape the same sort of scientific attention. The cultural etiology of mental illness renders it a complex object that maps onto complex subjectivities.

In addition to the slowing pipeline of new therapies, which many frame as a problem of scientific innovation (Insel 2009), industry insiders point to regulatory roadblocks and expiring patents (Miller 2010). The pathway between understanding a disease and creating a pharmaceutical solution is paved with myriad challenges. These challenges include a high failure rate of clinical trials, as well as the subjective nature of diagnosing and assessing, brain disorders (Ninnemann 2012).

However, for scientific explanations that point to the sheer complexity of the brain as the reason for the scientific problems, these deterministic narratives usually call for increased neuroscience knowledge in order to find therapeutic targets. Thus, TN, positioned as a solution to the problematic nature of commercial, for-profit neuroscience, promotes further engagement with psychopharmacological models of mental health. In this sense, the problems with old neuroscience are solvable with more neuroscience.

For scientific problem narratives (Fujimura 1996) founded on the complexity of the brain, the problem is the exceptional difficulty of finding *mechanisms of action* in the brain and biomarkers around which to develop therapeutics and approaches. While TN seeks to remedy this by supplying clinical scientists with increased basic knowledge of neuroscience and pathophysiology (FitzGerald 2005), issues such as depression and overeating do not neatly correspond to biomarkers in ways that are easily addressed solely through *pharmaceuticalization*. The problem of the social and the individual remains an unaddressed issue in the translational paradigm.

Many scholars have made the point that psychopharmaceutical solutions do not easily map onto given individual experience (Biehl 2005; Dumit 2003; Greenslit 2002; Jenkins 2011; Lock and Nguyen 2010; Ninnemann 2012; Schlosser and Ninnemann 2012; Whyte et al. 2003; local meanings and contexts (Das and Das 2006; Lovell 2007; Petryna and Kleinman 2006); biological difference and complexity (Lock and Nguyen 2010; Ninnemann 2012; Petryna 2006; Schlosser and Ninnemann 2012); or national and cultural narratives (Applbaum 2006; Ecks 2005; Good et al. 2007; Kirmayer and Raikhel 2009; Lock and Gordon 1988; Schlosser and Ninnemann 2012; Young 2007). In fact, the complexities at work in mental illness tend to point to the importance of understanding it as a product or *technophenomenon* (Lock and Nguyen 2010). Lock and Nguyen's notion of technophenomena addresses the way biomedicine's regimes necessarily construct illnesses as *technological problems*. Their unique reading of technological determinism illustrates the way technological framing can lend a very particular reality to the problem for which the technology was constructed as a solution; this is made especially evident in the approaches toward diagnostics and clinical measurement of complex phenomena such as depression.

According to scholars, the problems of measuring depression are part of a larger lack of understanding that has imperiled the science behind antidepressants: "People have not paid enough attention to how to measure depression, how to measure psychosis" (Van Gerven and Cohen 2011). The construction of mental illness as a brain disorder thus informs the conceptualization of a wide array of interventions, including the design of diagnostics. The large trend toward envisioning mental illness via the lens of genetics (Insel 2009) and the move by pharmaceutical companies to invest in genetic research programs in lieu of neuroscience foci (Abbott 2011; Bell 2013) may have given rise to a new kind of biotechnological conceptualization: *geneticization*[10] of brain-based medical disorders, something that is an emerging trend within TN and translational science in general.

Scholars such as Joseph Dumit have shown how the move toward defining disorders technologically—especially diagnostics and screening—dovetails with pharmaceutical companies' corporate strategies focused on commoditizing prevention, thereby pushing pharmaceuticalization on individuals who merely had the *risk* for a given disease. According to Dumit's (2012a) analysis, this trend reflects an emerging birth-to-death prescription model. By treating genetic predisposition, pharmaceutical companies move toward gene articulation (more specifically, genetic possibility) as a point of biomedical intervention, itself a veritable ground zero for twenty-first century global health response and a source of local, individual (Rouse 2009), and clinical (Keating and Cambrosio 2003; Fullwiley 2011) truth making.

However, in the case of clinical neuroscience, the move toward genetics as a new strategy may reflect, as suggested earlier, a failure of existing theories in neuroscience (Insel 2009), including existing models at work in research programs focused on depression, autism, and Alzheimer's disease. One finds a clear example of this sort of psychogeneticization of brain-based illnesses in Margaret Lock's (2013) work on Alzheimer's, in which she chronicles how a profound dearth of effective new solutions based on existing Alzheimer's research models led to a turn toward both prevention and gene-based approaches. Yet, as Lock suggests, the growing genetic reconceptualization of Alzheimer's has not entirely supplanted older hypotheses. She states, "The model is currently being questioned by an increasing number of key researchers, but has by no means been overturned, and continues to be a driving force, even as the entire Alzheimer enterprise moves to include prevention as a major goal" (2013, 52).

Given similar diagnostic and ontological problems regarding depression, biomedical logics that enable a view of depression as something conducive to measurement and standardization illustrate the multivalent expansion of epistemologies regarding mental health and suffering. These epistemologies may still be highly contested, but they still inform diagnostic categories such as those that one finds in the *Diagnostic and Statistical Manual* (DSM) and for which the NIMH's move to use Research Domain Criteria (RDoC) in lieu of the DSM was envisioned as a partial solution. Additionally, the scientific crisis of the antidepressant represents a literal problem of *translation*: translation being the imaginary bridge connecting the facts and realities of the laboratory with the world (Latour 1983). Science studies (à la Latour 1983, 1988; Knorr-Cetina 1981) commonly observe that laboratories exist in and create "alternate worlds," but these alternative realities may not have been operationalized in the pragmatic shaping of translation-focused laboratories of university campuses.

No matter what the laboratory setting, translating findings from a basic laboratory into the world includes a multitude of issues: epistemic, departmental, scientific, political, regulatory, and so forth. Translational science, which in the ideal model informs translational medicine, includes laboratory research focused on drug discovery. In the case of pharmaceuticals, it requires translation between brain sciences and pharmacological and human-drug interaction; in the case of brain devices, it requires an understanding of the bioengineering of devices attached to human bodies. Translational science ultimately aims to translate genomic (Addison 2017) and physiological data into systems of personalized medicine (Garcia-Rill 2012). Thus, translation gets at the way biomedicine suffers from an ever-evolving set of *multiple* mistranslations between, for example, pharmacological and individual bodies; between laboratory rats and clinical-test subjects; between genetic facts and their expression. Under the translational shift, translations can be multiple.

The ideology of TSM is, therefore, that it would be a remedy for the very real panoply of breakdowns that ultimately occur between the laboratory bench and the patient bedside—a process that necessarily sutures together wide-ranging fields, methods, interests, individual and personalized data, and institutional and social stakeholders—incorporating this heterogeneity to operate as a bridge.[11] In this sense, translation is as much a fantasy about integration as it is an anxiety about *falling apart*. Risks—especially scientific risks—are imbued with the very philosophy of TSM: replete with the underlying anxieties for which translation had been envisioned as a solution, brought to the fore, and displayed in narratives of scientific and clinical complexity in the aftermath of controversies over psychopharmaceutical efficacy and questionable gains in health.

## 3.9 Scientific Risk vs. Financial Risk: Knowledge Problems Are Not Always Financial Problems

The massive financial risks of CNS development eventually proved to be too much for many biopharmaceutical companies (Cutler and Riordan 2011). The financial risk in CNS included the exceptionally long development process, investor preference for short turnaround time (FitzGerald

2005), high failure rates of clinical trials, high risks and costs of regulatory hurdles, expiration of patents, and the growth of cheap generics. However, CNS had always been risky for investors as well as for biopharmaceutical companies. The knowledge problems that undergirded psychopharmaceuticals had been recognized throughout twentieth-century research priorities (Insel 2009), derived from a consistent pattern of failed clinical trials, speculation about efficacy, and complaints about R&D pipeline problems. But only more recently had CNS investments been considered financially risky. This reality produces the inevitable question about why investments in commercial neuroscience continued into the twenty-first century. How does a sector so prone to scientific uncertainty still attract massive private investment? Lastly, what happened to turn TSM into an opportunity for newly risk-weary biopharmaceutical companies?

I wanted to better understand the relationship between scientific and financial risks in CNS development. In May 2011, as a fellow at UC Berkeley's Center for Science, Technology, Medicine & Society, I was introduced to David Grosof, a neuroscientist who had worked as an academic at UC Berkeley. Like many scientists in the neurotechnology community, he had dual degrees: a PhD from UC Berkeley and an MBA from MIT. Grosof later returned to UC Berkeley to work in computational neuroscience, developing mathematical models from neurology to apply to commercial endeavors, eventually leaving to pursue an ophthalmological startup at Washington University in St. Louis, Missouri. His oscillation between academia and industry was typical among academics involved with the neurotechnology industry, creating new startups, and serving as experts for university-industry institutes.

However, Grosof was also a self-described admirer of Durkheim and, as is often the case with bohemian, modern scientists (Sunder Rajan 2006), he was familiar with anthropologists who studied scientists. In our first encounter, I was reminded of the reflexive nature of modern scientists who are deeply aware of their social worlds (Luhmann 1993). During our first conversation, he boasted of conversations with Paul Rabinow[12] and told me, "Chances are I've got more of an appetite for the hypotheses and emphases you're bringing to bear than most life scientists." Grosof, having experienced both academic and entrepreneurial contexts, offered insight into the industry-academic partnerships in TN, one that is attentive to the problems of expectations engendered through translational fantasies accorded to neuroscience:

Alzheimer's research is quite protean like cancer research in the 1970s ... yet the economic and political demand for a better treatment is huge as the aging U.S., European, and Japanese populations are creating a crisis in human and fiscal cost. Quite risky clinical trials of lousy drugs (which all failed so far) have been done because the existing drugs are so poor that any quarter-assed [the result of only the most minimal effort] treatment that's any better will get sales of a billion dollars a year. (Interview 22, March 23, 2011, Berkeley, California)

Investment risk inherent in certain neuroscience projects could be offset by the large market opportunity that neurotechnologies offer—a principal consideration in the evaluation of investments. The possibility of even incremental improvements in treatment would justify the high risks of CNS research and development.[13] Thus, market size is a principal element in biopharmaceutical companies' sustained interest in the sector, made evident by many companies' innovation *externalization*, rather than complete abandonment. According to NIMH, one in four adults suffered from a diagnosable mental disorder in any given year.[14] The gargantuan market for CNS made neuroscience a persistent investment target. Because mental health presented a massive opportunity, CNS—despite the knowledge problems on which it depended—offered the possibility of incomparable financial returns.

This lucrative investment opportunity has also been reflected in investing activity: enormous amounts of capital were invested in brain-based biotechnology companies between 2006 and 2009. Even in 2008, at the tail of the global financial crisis, venture investment in neuropharmaceutical companies totaled $977 billion (68% of the total neurotechnology investment for that year) and in brain-based device companies $390 million (27% of the total neurotechnology investment). Yet investments in neuropharmaceuticals and neurodiagnostics declined 25% compared to 2007 (NeuroInsights 2010), representing disinvestment from CNS. The high investment rates prior to 2008 occurred because CNS and its objects—antidepressants, diagnostics, peripheral technologies, and procedures—proved highly profitable (see figure 3.7). Nevertheless, active investment in CNS continued in 2010.

The $80.5 billion sales in 2010 showed CNS profitability despite skepticism regarding the science of psychopharmaceuticals and growing research about efficacy problems. Considering all neurotechnologies—diagnostics, surgical methods, devices, and services—2009 revenues were $141 billion,

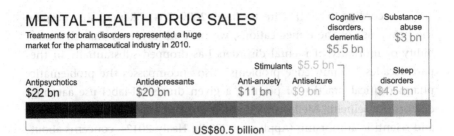

**Figure 3.7**

Sales from CNS in 2010 according to BCC Research (Abbott 2011)

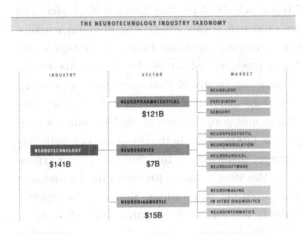

**Figure 3.8**

Neurotechnology industry taxonomy of each area's revenues (*The Neurotechnology Industry 2009 Report*, courtesy of Zack Lynch, NeuroInsights)

according to the *Neurotechnology Industry 2009 Report* (see figure 3.8). The enormous profitability of CNS products was also expected only to grow as psych diagnoses continued to rise alongside an ever-widening umbrella of what would be considered a psychiatric or neurological disorder. By 2016, the market for neuropharmaceuticals (pharmaceuticals to treat brain disorders, including psychiatric illnesses) had grown to $142 billion.

All of this emerged despite the accumulation of evidence that there are massive scientific problems at work in the theories that undergird psychopharmaceuticals and internal awareness of knowledge problems attached to pharmaceuticals (Healy 2012). "Knowledge problems" refers to the crises over efficacy (Fournier et al. 2010; Insel 2009; Ninnemann

2012). According to NIMH Director Insel (2009, 704), "Remarkably, despite the heavy use of these medications, we have no evidence that the morbidity or mortality of mental disorders has dropped substantially in the past decades." "Knowledge problems" also encompasses the problematic pharmaceutical practice of pushing a given drug's off-label use aamong doctors (Kesselheim, Mello, and Studdert 2011), problems of ghostwriting and scientific authorship (Applbaum 2009b; Healy 2012), concerns about massaged clinical trials (Petryna 2005), pharmaceutical truth constructions (Dumit 2012a; Joyce 2005, 2008; Matheson 2008), and problematic biological conflations (Ninnemann 2012). Also included in the category of "knowledge problems" is research about psychopharmaceutical safety and attempts to manage such research (Healy 2012). The question remains: How can a pharmaceutical company maintain high profit margins with a lack of novel products and products that contain knowledge problems?

In the essay "Is Marketing the Enemy of Pharmaceutical Innovation," Kalman Applbaum (2009c) suggested that *pharmaceutical value* had increasingly become a marketing proposition, bringing much R&D into the marketing process. For Applbaum (2009c, 15), "In the integration of marketing and R&D across contemporary medicine, we find fewer and fewer expressions of scientific value." In this view, the push for profits means a reliance on marketing, rather than scientific, innovations. Similarly, the pharmaceutical strategy to treat risk factors for diseases in anticipation of their possible expression (Dumit 2012b) illustrates a strategic focus on prescription maximization beyond sheer symptomology.

While Applbaum's (2009c) analysis relies on scientific optimism about cures culled from "honestly pursued scientific outcomes," pharmaceutical companies' practice of growing profits from marketing practices (Moncrieff 2006) rather than from therapeutic benefit has been especially well documented (Applbaum 2009c; Dumit 2012; FitzGerald 2005; Greenslit 2002). However, this also brings into view that while psychopharmaceuticals are laden with scientific or epistemic risks, there are still opportunities for positive financial returns. In other words, just because CNS relied on knowledge problems did not mean that their pursuits were not profitable.

The 2010 Neurotechnology Partnering and Investing Conference made this especially clear. Manuel Lopez-Figueroa, a vice president at Bay City Capital (a venture capital firm), discussed the status of psychiatry. He said, "I want to begin with the status of the psychiatry market. As you all know,

in the [next] couple of years, the number of drugs that will expire is significant, and it's about $95 billion—$95 billion will be expiring in cumulative
numbers." He then put up a slide that showed CNS had the lowest clinical
success. Another slide showed that CNS mergers and acquisitions had the
second highest level of transactions among all CNS areas.

The accumulation of too much financial risk by biopharmaceutical companies and the subsequent need to expel that risk is its own kind of *overleveraging*. Externalization is one mechanism by which biopharmaceutical
companies have dealt with the overleveraging and overcommodification
of risky products. Thus, the move toward the university as a space of externalization is a solution to this larger risk crisis. Only in the coming to light
of the massive scientific ambiguity folded into pharmacological solutions
and the acceleration of commercialization despite this ambiguity can we
understand pharmaceuticals as both epistemic and financial risk objects.

Thinking about the things that objects are composed of (Schüll 2010),
the social and technical assemblages (Pinch and Bijker 1984) contained
within biotechnologies, one could argue that neurotechnologies—pills,
devices, diagnostics—essentially "contain" those risks. Both financial and
scientific risk become part of the materiality of objects produced from neurotechnological R&D. Consumers of psychopharmaceuticals *ingest risk*. For
some, the very digestibility of risk reflects neoliberalism's impact—an issue
made clear in the case of the global financial crisis and public ingestion of
the risks from collateralized debt products. Returning to the question of
what psychopharmaceuticals are made of, I propose that they consist of a
panoply of scientific, knowledge, and health risks (Light, Lexchin, and Darrow 2013); the concretization of strategies to promote prescriptions based
on unclear evidence about their efficacy; and swelling investor and financial risk-taking. I argue that these materialities (or perhaps immaterialities)
buried inside psychopharmaceutical products are as significant as the pharmacological agents they also contain.

### 3.10   The University Is Not Merely Entrepreneurial

After the Clinical and Translational Science Award (CTSA) Program launch
in 2006, the institutional effect was sudden and dramatic. For the approxi-
mately 60 academic medical institutions across the country that launched
translational science centers, it was a collective and tectonic movement.

Since the CTSA program led to creating university-based centers and consortia, the university became a key player in institutionalizing the translational shift within neuroscience. This institutionalization also necessarily included reorganizing university departments and programs, rearranging institutional missions and priorities, and injecting a translational ethos into existing departments. According to the CTSA program description: "CTSA institutions provide core resources, essential mentoring and training, and opportunities to develop innovative approaches and technologies designed to re-engineer existing translational research capabilities. By providing integrated academic homes for clinical and translational science, these awards foster change" (Clinical and Translational Science Awards n.d.).

Thus, translational science becomes not merely additive or extensional; it seeks to reengineer and supplant existing academic organizational landscapes. This is about institutional transformation. The genesis for these new institutional forms began around 2005 when a host of stakeholders were thinking about how to institutionally rearticulate translational programs at universities.

University research programs were also redesigned under the interdisciplinary aegis. The ideology of translational science proposed that disciplinarity and academic silos were primary culprits in the problems that translation sought to fix (FitzGerald 2005; Garcia-Rill 2012; Klein 1996). In nearly all definitions of science and medicine—whether articulated in the *NIH Roadmap* or by pharmaceutical companies touting their innovations—there was a strong emphasis on creating partnerships *across* fields, disciplines, methods, and models. Perhaps translational science was obsessed with cross-disciplinarity, but only inasmuch as this cross-disciplinarity afforded instrumentalism—solving *particular* problems, focusing on *particular* agendas. In other words, undoing barriers was about a methodological and conceptual strategy to facilitate cross-disciplinary productivity and create shared values and agendas.

One way to read this is that a shift toward the latter—a collective and shared ethos important for commercialization—became a working ethos at the university. This reading allows one to see TSM as being about designing an ethos to enable commercialization, producing operational systems to create cleaner pathways and reduced barriers to collaboration, and creating technical strategies to generate greater operational efficiencies— all toward the goal of "efficient commercialization of their intellectual property" (FitzGerald 2005, 817). Recounting this short institutional trajectory of translational science connects to thinking about the role of externalization as a means by which the university is brought into commercial partnership with biopharmaceutical companies and allows a renewed understanding of the relationship between TN, translational medicine, and health.

I contend that the turn to the university—that is, the focus on the university as a critical enactor of translational and translation-focused neuroscience and as the space *charged* with neuroscience innovation—should not be taken merely as a move toward better neuroscience or by extension better health. Rather, this institutional emergence is notably about enunciating a new role for the research university amid a particular moment in healthcare corporatization and the effects of the fallout from the financial risks of over-leveraging the financial and medical promises of psychopharmaceuticals.

Thus, we can think about externalization as a means by which (1) university-based TN provides financial opportunity for private firms through offloading risk during a moment of crisis for the sector, (2) academic medical centers can commercialize their underleveraged assets in the form of intellectual property and become a new class of "vendors" for a newly "vertically disintegrated" pharmaceutical sector, (3) corporate firms and investors can access a specialized route to the privatization of public resources, and thus (4) TN becomes a strategic solution for the problems brought about by the financialization of global biomedicine.

The "entrepreneurial university" is indeed at work. Scholars have analyzed and documented many ways in which universities became implicated in rationalization and bureaucracy (Apple 2005; Krimsky 2004; Shore and Wright 1999; Strathern 2000); turned toward the market (Berman 2012; Bok 2003; Buchbinder 1993; Hong and Walsh 2009; Olssen and Peters 2005; Orr 1997; Shumar 1997); underwent corporatization (Etzkowitz et al. 2000; Readings 1997; Schreker 2010) and privatization (Brown 2000; Lave, Mirowski, and Randalls 2010); and bifurcated into university-industry segments of large-scale research firms (Etzkowitz 2003; Lam 2010; Lowen 1997). The trajectory that rendered the university a key player in economic development and regional and global innovation is long and complex (Apple 2005; Berman 2012; Mowery et al. 2004) and university transformational scholarship long-standing (Brett 1945; Veblen 1918). In fact, Vannevar Bush, a founder of the National Science Foundation, had always included the privatization of public funding of science research in his vision for innovation in the US. However, recent work tends to narrate the relationship between the life sciences and capitalism as being predominantly about capitalism's logics at work in both. That view is, at best, incomplete.

One cannot appropriately contextualize the rise of TSM at the university without also accounting for the broader trajectories of which the university is part: the move to push university-based scientific discoveries toward commercialization, reflecting an entrepreneurial or startup culture at the university (Etzkowitz 2003; Sunder Rajan 2012); the demands placed on knowledge workers in capitalist societies (Shapin 2008); and the transformation of the modern university. Accordingly, much of the scholarship focused on university transformation has brought into view the ineluctable role of neoliberalism (Cooper 2008; Lave, Mirowski, and Randalls 2010; Olssen and Peters 2005; Shore and Wright 1999; Strathern 2000). The commodification of university activity, the privatization push at the university with respect to licensing and patenting intellectual property, and the defunding of public universities all speak to the logics of neoliberalism at work.

Yet scholars have suggested that science and technology studies have not adequately addressed or theorized "the deeper changes in scientific practice, management, and content as neoliberal concepts have been used to justify major innovations in the structure and organization of science" (Lave, Mirowski, and Randalls 2010, 644). In this sense, the translational shift in university neuroscience, and TSM in general, must be seen as evidence of a much larger change in the organization and systematization of *knowledge in general* under neoliberal ideology.[15]

More recently, explicit investigations of the interplay between the life sciences and capitalism have also delved into how universities have played a central role in the commercial trajectories of the life sciences (Berman 2012; Cooper 2008; Haraway 1997; Herder 2013; Kleinman 2003; Mirowski 2011; Sunder Rajan 2017). In work recounting the trajectory of biotechnology commercialization at Stanford, Sunder Rajan (2012, 4) suggested that the university has been implicated in the "changing nature and locus of 'commercialization' … with the university becoming a more explicit- and institutionally-regulated stakeholder in the entire process." Thus, few would argue against the contention that economic logics have animated both the life sciences and the university in a "neoliberal direction" or that increasing university privatization toward the market has affected university practices and revised their central aims.

However, for the life sciences, the shift at the university is only part of the overall picture. Rather than a broad-based theorization of capitalism's transformation or triumphalist narratives that see TN as merely a manifestation of neuroscience's advances, there is space for viewing translational science and TN functions in relation to specific financial realities that affected the biotechnology sector during 2006–2010. During those years, concrete financial situations occurred in the global economy—situations that presented particular consequences for pharmaceutical and life science innovation. In this sense, there is an opportunity to see the political economic formulations that resulted from the chaos and degeneration of psychopharmaceutical promises. It is imperative to retain a view of TN operating in the context of a moment of knowledge *failure as such* (Miyazaki and Riles 2005)—in this case, the massive problems and crises drug development and brain-based biotechnology sectors faced.

In short, the translational ethos was not merely neoliberalism at the university. Rather, it was a coordinated and strategic transfer of the risks of early-stage research and investment from the private to the public sector, as well as the biotechnology sector's articulation of innovation strategies created in anticipation of impending financial losses produced via speculative pharmaceutical markets.[16] Clearly, to the extent to which TSM enabled the driving down of the costs and risk of R&D and in doing so benefited shareholder value, TN and TSM in these ways function as a financial strategy even before any "translation" even happens. Together, the observations about what externalization did and how it functioned lead to the argument that rather than being a form of capitalism—an argument in line with current theorizations—TN functions, more precisely, as a form of finance.

## 3.11 Vertical Disintegration

My analysis suggests that TN, and TSM more generally, "must be understood—at least in one sense—as something that specific companies sought out with an eye toward diverting the costs of early-stage R&D onto the public domain, a fact that is only visible through an analysis of the on-the-ground articulations of TN, including its emergent partnerships as well as those partners' corporate histories and reconfigurations" (Robinson 2018, 249).

However, additional research shows that the corporate partners of TN used TN not only to offload costs, but also to externalize risks from early-stage R&D. The riskiness of executing high-stakes scientific projects within companies, which became especially anxiety producing for investors during the global financial crisis, became part of the justification for vertical disintegration—a term used to describe sectorwide trends toward externalizing or outsourcing multiple parts of a company's formerly internal business activities. These activities might include publishing companies' outsourcing of content creators and writers or the decisions of car manufacturers to use Chinese manufacturers to create more and more parts and systems for their automobiles. A product's entire supply chain (whether product development, manufacturing, etc.) becomes potentially externalizable under vertical disintegration. Vertical disintegration describes two sector processes co-occurring. When companies begin to outsource and offshore many parts of their production system, the process, by its very constitution, produces entire new markets and industries consisting of specialized vendors especially suited to execute the now-outsourced labor and projects created as excess. In some ways, one can envision TN as a new industrial market—part of "the emergence of new intermediate markets that divide a previously integrated production process" (Jacobides 2005, 465). In this way, TN is both part of a system of externalization and the product of the biopharmaceutical sector's reconfiguration—a fact made clear through the spate of pharmaceutical companies restructuring from 2005 onward. To analyze the history of TN without accounting for the role of the sector's move toward externalization—of which vertical disintegration is but one example—is to misunderstand the political and economic context out of which TN emerged.

As universities embark on their own pharmaceutical and biotechnological development, and as they begin to openly acknowledge "drug discovery" as part of their academic mission, the "bridge" represented by TN produces questions about university workers—for example, whether the priorities of funders and the financial opportunities TN entails affect not only the priorities of research but also the ways those who carry it out understand their particular research culturally and ethically. TN exemplifies the growing entanglement between academic research and market-oriented approaches toward knowledge production (Lauto and Valentin 2016; Power 1999; Strathern 2000)—something discussed in much of the knowledge and privatization literature as well as literature focused on the effects of neoliberalism on scientific production. Clearly, much modern biological research is becoming "translational" in this sense. Neuroscience, as one of the newest fields in the life sciences and despite its outcome failures, is ahead of the curve in its self-consciousness about its translational imperative.[17] TN can be a model for the translational shift—both because of the mystery surrounding the brain and because of the work pharmaceutical companies are doing to craft markets into which to bring neuroscience innovations.

But the narrative of TN as an inevitable evolutionary advance in the history of neuroscience causes obfuscation. Like many of neoliberalism's revisions, this historiography hides the story of a system falling apart. However, a large number of recent life science theorizations tend to treat the manifestations of neoliberalism in ways that presume that they were wholly successful. Grand theorizations about the emergences and evolution of capitalism within the life sciences may inadvertently have absolved capitalism of its failures and treated it merely as a genesis—one more development of capitalism's many intrusions. Mapping internal corporate histories, histories of failed scientific models, as well as specific sector shifts—such as those that occurred around neuroscience R&D for Big Pharma—combined with ethnographic attention to on-the-ground transformations and critical stakeholder perspectives provide the needed evidence from which such failures become clear. Contemporary studies of neoliberalism's formations and malformations must also include a thanatological imperative—to document when certain economic models break down—as well as an ethnographic specificity that does not miss the limits, messes, explicit failures, and subsequent debunking of economic theories and their formations (Miyazaki and Riles 2005), which, as I argue in the sections and chapters that follow, exemplify the story of TN and TSM writ large. Several sets of intersecting failures are a crucial part of the story of the translational rise.

The rehabilitative needs of financialization played a vital role in the emergence of TN at universities and represented the amalgamation of a new form of neuroscience innovation created through the consequences of overleveraged psychopharmacology markets. It was not only a transfer of research portfolios and corporate R&D logics, but also a transfer of risk and subsequent obligations—now accorded to the university and other stakeholders—of de-risking. In other words, the falling down of CNS created a particular de-risking function for this form of university neuroscience research—a fact that would have wide-ranging implications for patients and research.

At the same time, TN is about much more than a specific history of healthcare corporatization and its financial functionality. As Zack Lynch, the prominent "Big Man" (an important person or figure in anthropological parlance) of neurotechnology whom I introduced at the outset forcefully suggested, TN and translation, in general, occupy a space of envisioning. For Zack, TN was the key to a gilded future and a means for improving the human condition. In fact, this market envisioning is just one of multiple fantasies that the translational imagination enables. In the chapters that follow, I ask: what are the means, avenues, and systems through which financialization affects university knowledge processes? Later I consider implications of translational science and medicine's constraints on the scientific imagination and their potential impact on patients.

# 4 The Bench: Universities and Laboratories under Translation

One question that emerges in the study of translation is how the work done in translational laboratories is understood as innovation in health. This chapter shows how the scientific laboratory operates as a critical site in which such suturing occurs. In what follows, I outline how the design of environments, especially laboratories, software, and university research parks, enables an important vehicle through which translational work becomes both connected to—and understood as meaningful in relation to—the category of health. Through an examination of discourse as well as environments such as architecture and software, we see the many material means through which translational research becomes *constituted* as an innovation in health itself. Through an examination of built environments, one also unearths essential practical mechanics: how much of the work of translation involves connecting, partnering, and assembling translational activity to biomedicine and its accoutrements. In tracing how translational research becomes biomedically and economically meaningful, we also unearth the many moments of "translation" that suture objects, activities, facts, and narratives to patients, medicines, therapeutic conceptualization, and the larger project of health. We also see the myriad "architectures" that enable the translational shift to connect science and scientists to capital.

## 4.1 Open Questions about Epistemologies and Environments: The Importance of Design

As universities are uniquely implicated in TSM (Garcia-Rill 2012; Kurtzman 2018; Pfotenhauer and Jasanoff 2017), its role in relation to particular impacts of neoliberalism on academic science is an essential area for investigation—especially with the increasing relevance of and expectations for commercialization of university scientific work (Etzkowitz 2003; Krimsky 2004; Herder 2013; Lam 2010). Despite the growing relevance of this issue to the social study of science, there is a dearth of scholarship especially in science and technology studies about the specific mechanics and particularities of scientific practices and knowledge systems affected by neoliberalism and pressures toward commodification (Lave, Mirowski, and Randalls 2010; Sismondo 2009). In this section, I trace how various levers help to shape TN, showing how spaces—both internal and external to the traditional scientific laboratory—ultimately impact and shape the kinds of science done within these laboratories. Ultimately, these spaces reflect designs for a translational mode of knowledge—in which *all research* becomes potentially translatable, or at least pretranslational. Accordingly, these designs also reflect and imbue desires for specific kinds of scientific and intellectual outcomes—outcomes, for example, made explicit at Neu-rotechnology Investing and Partnering Conferences where investors and pharmaceutical executives explicitly list the kinds of projects in which they have investment interest in front of an audience of university scientists and biotechnology startup founders.

Recent research that attempted to map the collocation of financial logics within the new epistemologies of the life sciences performed a critical sim-plification. For example, AstraZeneca's decision to create, plan, and design virtual innovation spaces structured primarily around academic partners, and the decision by the University of California at San Francisco (UCSF) to help design and construct a new physical neighborhood into which pharmaceutical companies could relocate their R&D arms, were not merely products of the life sciences' institutionalization, coevolving financial and scientific structures, or sheer epistemic transformation. These trans-formations reflect a specific set of causes brought about by the increasing financialization of biomedical science and shifts that co-occurred along-side healthcare globalization. Clearly, the translational shift and life science financialization were poised to affect epistemologies (Ecks 2008; Sunder Rajan and Leonelli 2013). However, there is a real need to map precisely how these new formulations created durable epistemic effects, as well as the ethical impulses at work in the task of design.

This consideration of how systems and environments help to enable TN also brings to view the continued importance of studying infrastruc-tures—an issue that has brought about renewed scholarly and emergent anthropological reflections on the ways that infrastructures are both "connective" forms and "material" ones (Venkatesan et al. 2018). The various ways that scholars have considered infrastructures (Star and Ruhleder 1996)—as concretizations of ideologies, means of creating and sustaining interpersonal connections, material embodiments of politics—are found here in the case of TN and the massive set of environments and structures built to engender and sustain translational logics.

## 4.2  Liquefaction: A Short History of Mission Bay

My 2010 fieldwork consisted of observing in a Northern California university laboratory, as well as intense involvement with the larger neurotechnology industry community. In addition, I became part of San Francisco's biotechnology startup and entrepreneurial communities located in the newly developed neighborhood of Mission Bay (Timmerman 2010), which was home to UCSF's new biomedical research campus. UCSF is a highly ranked medical center and teaching hospital known for its entrepreneurial ethos, research productivity, and connections to Silicon Valley.

A long-neglected industrial neighborhood in San Francisco, Mission Bay is located in the eastern portion of the city. The area's adjacency to the San Francisco Bay meant that historically it had been an important space for industrial and commercial work. In the mid- to late nineteenth century (see figure 4.1), Mission Bay was home to shipping companies that used the bay as a dumping site as well as to slaughterhouses that would send blood and guts from slaughtered animals directly into Mission Creek to float out into the bay (Conor 2007). In 2000, the city embarked on a massive cleanup effort to remedy the area's toxic soil after years of industrial pollution. After San Francisco designated the area as a redevelopment zone in 1998, Mission Bay morphed into a new, strange geography that seemed suddenly to emerge from the ground (Allday 2013). By 2003, UCSF had built a 34-acre campus in the zone. In 2004, two life science companies moved to the area. By 2012, over 54 life science companies along with a smattering of other research institutes, companies, and venture capital firms (Ross 2013) called Mission Bay home (see figure 4.2). Once home to railroads, warehouses, dilapidated buildings, festering shipyards, and small communities for those industries' workers, the Mission Bay neighborhood looks and is entirely different today.

**Figure 4.1**
Rendering of Mission Bay and the former Mission Creek in the 1870s (@Library of Congress)

**Figure 4.2**
Mission Bay in 2009. *Photo: Chris Carlsson.* Attribution-Noncommercial-Share Alike 3.0. *Courtesy of FoundSF.*

Mission Bay is topographically and geographically unique. Compared to San Francisco's steep hills, tall trees, and Victorian buildings, Mission Bay's eerily flat terrain and contemporary architecture seem removed and peculiar, producing the feeling of an entirely imported neighborhood. The strangeness is ubiquitous. Walking around Mission Bay, one can see construction everywhere. Entire landscapes and city blocks emerge quickly. Fully mature trees are brought in from elsewhere. Although different, the erected buildings feel aesthetically and textually similar: glassy, contemporary, and thematic. Even in their faux difference, or what has been called "demassified difference" (Nye 2007, 610), the architectural similarity is more prominent than any aesthetic novelty.

UCSF was a significant stakeholder in Mission Bay's massive development. Located prominently in the center of the new campus is its centerpiece, Genentech Hall (see figure 4.3), named after the Bay Area biotechnology company started at UCSF. The sprawling new campus contains new laboratories, academic buildings, dormitories, and a new hospital, which opened in 2015. In 2017, UCSF announced that a new building would be built as the headquarters for the UCSF Weill Institute for Neurosciences. The Mission Bay neighborhood contains conference centers, hotels, and corporate office parks. Mission Bay also came to represent the verdant and shimmering youth extolled in narratives of Northern California innovation. Lured by a flurry of controversial tax incentives,[1] starting in 2011 several biotechnology companies relocated to San Francisco from other states, some at the suggestion of largely California-based venture capital firms that wanted to be close to their portfolio of companies.

**Figure 4.3**
Genentech Hall. Photo: Michael Macor, *The Chronicle*. Copyright © SF Gate.

In 2011, the German pharmaceutical and chemical company Bayer moved its R&D hub to Mission Bay, entering into a 10-year agreement to produce an industry-university partnership with UCSF. Adjacent to its and UCSF's R&D facilities, Bayer created a facility program called "the CoLaborator," an aptly named shared workspace for healthcare-oriented entrepreneurial startups. Among several healthcare and pharmaceutical companies opening up collaborative "innovation" spaces in Mission Bay (Ross 2013), Bayer's CoLaborator was part of an innovation strategy focused on exploiting partnerships. Local startups could also participate in a suite of services offered by Bayer and nearby UCSF. Indubitably, Bayer's narrative positioned these new models under the aegis of forging innovations in health:

> At Bayer, we believe collaboration is an essential way to bring new therapies to the patient. This philosophy led us to open the CoLaborator, a unique incubator space for start-up companies next door to our U.S. Innovation Center in the Mission Bay neighborhood of San Francisco, CA. Located in the heart of the city's thriving life sciences cluster, the CoLaborator is within walking distance of UCSF, QB3, the Gladstone Institutes, several venture capital groups, and more than 60 emerging life science companies, making the Mission Bay location well-suited to support the success of a start-up company. The CoLaborator's design is a flexible, open floor plan with 6,000 square feet of shared, rented lab space designed to house startup life science companies whose technology platforms, drug targets, or drug candidates may align with Bayer's interests. The environment fosters opportunities for idea exchange by bringing researchers together in a common space. The CoLaborator includes basic equipment for life science startups to quickly begin putting their ideas to the test. Bayer support includes Environmental Health & Safety and Biosafety licenses, and access to nearby UCSF core services such as imaging, bioinformatics, and proteomics. Partnering with Bayer can also provide access to the global expertise and equipment of Bayer's research network.[2]

Bayer's description illustrates that, as part of UCSF's development, Mission Bay became a regional and national hub for life science innovation; the neighborhood is now home to a peculiar mix from academia, young startups, and venture capital. This very mix was understood to be necessary to foster innovation (Pfotenhauer and Jasanoff 2017) and in particular, innovation in health under the translational imaginary. Nevertheless, as I wandered around Mission Bay staring at a sea of glass facades, building foundations, and truckbeds with fully grown palm trees awaiting installation, I wondered why this San Francisco neighborhood was so flat.

Late one afternoon, I noticed a man pushing a cart full of found objects and cans off 16th Street where Mission Bay meets the Potrero Hill neighborhood. I asked the man, "Why is that area over there so flat?" Behind us to the east, Potrero Hill's hilly terrain protruded into the evening sky. Gesturing west toward the flurry of new buildings, he responded, "It's all built on sand." He smiled and shrugged his shoulders. I learned that much of the neighborhood of Mission Bay was "filled in" with sand to enable building into areas that had once been part of the bay. The part of Mission Bay where the UCSF campus was built lies directly on top of this new ground.

According to one engineer, "It's like building on a bed of Jell-O" (Ginsberg 2008). In-filled areas are not rare in the United States. However, for San Francisco the problem is earthquakes. In areas composed of landfill, earthquakes cause the land to undergo what geologists called "liquefaction." In fact, UCSF's Mission Bay campus—a throng of new high-rise residential towers—and many of the biotechnology companies' buildings are located in a designated "liquefaction hazard zone" (Zeng, Hiltunen, and Manzari 2008, 4).[3] The irony was glaring: this new physical environment, an impromptu global center for innovation and itself an ecosystem of sorts, was constructed atop a bedrock of risk, both financial and environmental.

### 4.3   Proximity + Intimacy = Epistemology

To map the institutionalization of translational science at the university, I needed to understand how translational science was being "done" at the university—to sit with administrators helping to realize the translational visions outlined by the NIH and trace their effects inside the laboratory itself. I thought that to quantify the translational shift's manifestations in the laboratory, I would need to spend most of my time inside the laboratory. However, spending time at the university showed how much of the translation work was about the broader environments of which the laboratory was part. Rather than the laboratory undergoing change, the laboratory was brought into an emerging configuration: a larger system through which the work of translation occurred. The issue at hand is the means through which the university was implicated in a changing system: a change discernible through attention to changing networks, systems, articulations, formations, and technologies that enabled institutionalizing university-industry partnerships and the shift in university research toward translational aims. Through this lens, the material ways these proximities affect epistemologies are revealed. One of the key means of creating such proximities has been in the design of university research campuses.

The emergence of the Mission Bay campus at UCSF was an opportunity to explore translational research as it emerged as part of a campus design effort. However, I soon learned that the "environments" designed to compel translational activity included both physical and digital structures. In September 2010, in conversation with an administrator at the UCSF Clinical and Translational Science Institute, I raised a question about the institute's location: I just could not figure it out. I had found the institute's website and searched several digital maps for information about its location. However, I had difficulty finding an actual building that housed the center on the UCSF campus. It was almost as if the institute were invisible—a fiction. I eventually found the building that housed the center's administrative offices, but this was not where translation "happened," as I had originally assumed. I would later discover that much of the work of "translation" that occurred via UCSF was managed through a web-based collaboration process. In some ways, the primary center was a digital one.

An office building on the far northeast side of Mission Bay housed the center at some remove from the UCSF's other campus buildings. The sign for the institute pointed toward a modern urban office complex called China Basin. China Basin was the complex that the drug company Pfizer had initially considered as a place to house its new Center for Therapeutic Innovation focused on creating new biotechnologies, but Pfizer had backed out of that location.

Pfizer had entered into a strategic partnership with UCSF. In 2013, UCSF provided a press release about this collaboration:

> The collaboration builds upon an October 2010 agreement between CTI [Pfizer's Center for Therapeutic Innovation] and UCSF in large-molecule—also called biologics—discovery, with both agreements aiming to accelerate the translation of biomedical research into new medications and therapies. According to the terms of this new collaboration, CTI will provide UCSF with funding and scientific expertise. ... If drug candidates achieve agreed-upon objectives, the partnership between CTI and UCSF could move from discovery into clinical trials, leading to potential milestone payments and royalties in the event a product is commercialized. (Bole 2011)

Pfizer eventually settled on a location *closer* to the broader UCSF campus. Proximity was part of a larger strategy through which the university, via translational initiatives and transformations, was becoming the biopharmaceutical companies' research arm. "Proximity" does not refer only to physical proximity but maps a panoply of intimacies achieved through the creation of connections: agreements, joint ventures, technologies, as well as physical relocations. In the articulation above, one sees quite literally the means, terms, contingencies, timelines, and structures (financial and legal) that underlay and enabled this proximity.

The mechanics of handing off to universities specific biopharmaceutical projects and research agendas, and how these transfers functioned as a corporate strategy, requires understanding that this relationship was not merely an extension of Pfizer's existing research infrastructure. Instead, it represented Pfizer's transfer to university partners of research portfolios that had been internal and the implications of this transfer. By 2011, Pfizer had reduced its R&D infrastructure by 35% (Jarvis 2012), laying off employees and shutting down laboratories and manufacturing centers. In 2012, Pfizer opened up its Centers for Therapeutic Innovation (CTIs), starting in San Francisco, then in Boston, San Diego, and New York City (Said 2011). Pfizer's was part of a cascade of layoffs among a host of pharmaceutical companies (Reisch 2011).

One common factor in many pharmaceutical-company layoffs between 2010 and 2013 was that the brunt of the reductions fell, in the words of a Novartis vice president for research and development spoken during an investor phone call, "most heavily on scientists in early discovery" (Reisch 2011). For Pfizer, university partnerships represented both an internal strategy and a means of dealing with its own risk. Thus, proximity became part of a bifurcation process whereby university translational research programs hosted initial-phase research that corporations found too burdensome or risky. Partner-university laboratories were suddenly populated with new and specific projects, agendas, corporate interlocutors, and funding sources. A clearer example of biopharmaceutical companies handing off their own early-stage research to university laboratories is the NIH initiative mentioned earlier, Discovering New Therapeutic Uses for Existing Molecules. In this little-known program piloted in 2012, the NIH brokered relationships between pharmaceutical companies and academic researchers, whereby each university took over a commercially failed research program on a single molecular compound.

It is important to note that pharmaceutical companies ceased research (often part of a general cost-cutting directive,) and passed on the project and its imperatives, presumptions, and risks to the research university domain. It is in this research transfer from pharmaceutical to university laboratories that we see the morphing of university translational laboratories into de facto biopharmaceutical-company laboratories. This leads to questions of the unseen risks to the public in the transfers, transformations, and rearrangements that occurred by enacting novel public-private partnerships touted as necessary solutions in forging innovation (Horvath 2004). Handing off to the university early-stage research also creates questions about the implications and costs to the now-"translational" university for taking on projects that tend to have a 95% failure rate (National Center for Advancing Translational Sciences 2013) and in a context of relatively few highly profitable university licensing revenues. Since pharmaceutical companies such as Pfizer and Novartis were able originally to ingest these costs through reaping billions of dollars in profit, what does it mean if these costs are externalized onto universities that lack the same profit component? Lastly, what does objectivity mean in the context in which a pharmaceutical company has outlined research objectives for a partnering academic laboratory and also (as in the case of Pfizer's CTI programs) funds laboratory researchers and PhD students?

Additionally, the NIH's brokering of relationships between pharmaceutical companies and US universities brings into view questions about how translational science works in geospatial rearrangements. It was in the creation of national and international partnerships between university research and pharmaceutical scientists where one could see the importance of proximities: the creation of environments. Thus, it was no surprise that Pfizer's CTI also relied on incorporating universities and creating proximity. Since the creation of its program, Pfizer has placed each CTI very close or even adjacent to a prominent research university. This adjacency is no accident. In a journal article focused on CTIs, one sees the key ingredients of the collaboration between Pfizer scientists—brought in specifically to work in these new CTIs—and academic scientists of the adjacent university:

> Pfizer and academic scientists appear to work side by side on a weekly, if not daily, basis, trading reagents, sharing equipment, and exchanging data and knowledge. "What's so cool is they have badge access here," Martinez says of his academic collaborators. "That proximity really fosters a quick exchange and ease of

interaction." The same is true at Pfizer's other CTI sites. "The CTI guys have UCSF IDs, and the UCSF guys have Pfizer IDs," Coyle says. The access reinforces the notion "that this is a joint project team and we're doing it together." (Bole 2011)

The symbolism of keycard access is telling. Indeed, the imagery of openness, sharing, and togetherness is a hallmark of the often-romantic language of innovation. However, in this snippet one sees in explicit terms the enactment of a model of an externalization strategy coupled with a risk strategy that enabled Pfizer to discharge the risks of managing its own internal innovation strategy, but that did not preclude Pfizer from exercising options over exclusive rights to the commercialization of anything useful that might emerge from the research of its new partner lab.

The more compelling part of these articulations of partnerships is in how they reveal ways a partnership might necessarily structure the kinds of questions asked and the objectives created within the UCSF laboratory. The questions themselves become the domain of the translational imperative. Thus, Pfizer's proximity is also intellectual—here, proximity nearly becomes epistemology. Quite clearly, this extrapolation of a corporate-style R&D process at UCSF has implications for the kinds of scientific practices considered useful. The very notion of "agreed-upon objectives" on the part of UCSF and Pfizer suggests not merely imbrications, but an articulation of focused objectives.

Another seemingly minor but crucial rearrangement evolved between pharmaceutical companies and their university partners. As part of its design, the NIH's Discovering New Therapeutic Uses for Existing Molecules program created a tool to speed up the legal and intellectual property process between universities and pharmaceutical companies using so-called template agreements (National Center for Advancing Translational Sciences 2013). According to Christopher P. Austin, NCATS director, these template agreements "streamline the process by limiting the amount of negotiation required before a project can begin." This streamlining process achieved through the template agreement also points to the potential diminishing of negotiation over risk as an additional strategy within the translational-shift risk transfer.[4] The creation of template agreements (exemplified in the very notion of a ready-made template) constituted a system of reproduction through which complex legal barriers were surmounted more systematically, and in which *ways of thinking* or *not thinking* about collaboration became automated. In this sense, translational science's institutionalizations are especially epistemic. In other words, a template agreement is a crucial structure or system that produces and reproduces; it eases negotiation and thinking about new partnerships. That Pfizer placed CTIs adjacent to universities while closing its own research laboratories manifests the creation of public-private environments designed for specific knowledge purposes.

Another example of an environment that produced epistemic effects was the decision to have "agreed-upon objectives." Such objectives created *durable* epistemic effects in the everyday interactions between Pfizer and UCSF stakeholders. Templates and objectives speak to epistemic systematicity supported by material/environmental conditions. The development of Mission Bay (see figure 4.4) is perhaps the clearest example of an engineered environment: a grand, impromptu microworld, not merely a new hybrid environment in which pharmaceutical companies and university labs are placed side by side, but also a knowledge ecology that, importantly, is *rendered more stable* via its new geographic arrangements and proximities. This stability explains the translational imperative remains, even as financial stakes and scientific difficulty abounds.

I suggest that the new partnership between UCSF and Pfizer's CTI was not only about creating "commercialized products," but also about creating *conditions* in which such commercialization becomes easier, efficient, and prolific. It is this kind of "commercial architecture" (Robinson,

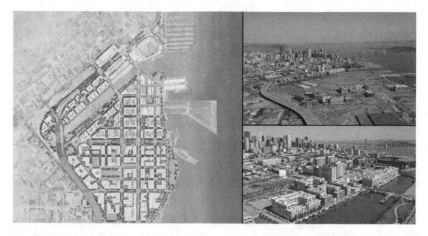

**Figure 4.4**
Left: San Francisco's new Mission Bay neighborhood. Right: Mission Bay's rapid and dramatic geographic transformation.

forthcoming, b) that makes TSM distinct from previous efforts at creating application-oriented modes of biomedical research. Infrastructures are vital to understanding the durability and potential longevity of translational paradigms.

## 4.4   The Search for Invisible Centers: Technologies of Collaboration and Facilitation

Inside the China Basin complex, a four-story complex of offices on the bank of a small river on the northern edge of Mission Bay, I located the administrative offices of UCSF's Clinical and Translational Science Institute (CTSI). Several floors up and far down a lonely hallway, the institute was surrounded by startups and other offices. Its location felt peripheral. On walking into the CTSI,[5] I noticed the space felt more like a minor waiting room than a collaboration or startup incubation space. There were few wide-open sites for active collaboration; instead, on one side of the office was a bank of administrative offices for the institute staff.

Placed prominently in the middle of the offices was a large conference room enclosed in glass. The expensive, high-tech chairs, large monitor, and sophisticated conferencing system visually dominated the room. The space itself suggested to me, perhaps presciently, that it was a meeting place for late-stage deal making rather than the open collaboration space that I had expected. The institute director, Maninder (Mini) Kahlon, welcomed me, shook my hand, led me into the conference room where she provided me with her card, and agreed to an interview regarding the overall structure and role of the CTSI. Kahlon's style was direct: "We're going to work with industry to find out what they need—and it's connected to industry's need and Pharma has found itself in a crisis right now, and the pipeline has dried [up], and they are even more eager to create partnerships."

She revealed CTSI's proactive approach to partnership. This proactive mode was a significant contrast to narratives in which translational science centers were simply waiting for industry to approach, or merely working on discovery-oriented neuroscience research. In this sense, the quality of autonomy or pure creativity associated with the concept of discovery felt like an inept descriptor of the process Kahlon described. The language of corporate "need" betrayed narratives of unbridled innovation and vocation attached to the translational science enterprise; this symbiosis gave university laboratories a particular orientation toward corporate functionality. And universities actively sought out these partnerships.

Kahlon's responses were quick and terse. Given its location in Northern California, "ground zero" for biotechnology startups, UCSF had a prolific history in terms of its partnering activities. Like its counterpart, Stanford University (Lowen 1997), UCSF had both a long history and an open stance that aggressively embraced the free market, encouraged faculty entrepreneurship, and did not shy away from explicit industry embeddedness. The new campus's physical location among a slew of biotechnology and venture capital firms was more than a metaphor for intimacies with industry.

Yet the growth of public-private partnerships was not merely about creating discrete programmatic and external relations between institutions. As anthropologists and historians have noted, objects are also transformed through connection. In the case of the university, this meant that the reconfigurations and transformations within universities were about systems making. In this way, there would be no need for UCSF's translation activities to be housed in some expansive physical building, or perhaps any building at all.

In searching the UCSF website to find the location of its translational science institute, I found very little information about the institute itself. Still interested in this issue of CTSI's location, I asked Kahlon, "Where *exactly is* the center? Is this it?" She caught on to my question:

> If you're thinking brick-and-mortar, this [where we're now sitting] is it. So the board [of directors] is here, and the central administration meets here, but then there are other programs, and they have leadership across campus—the physical site. But we emphasize the virtual site because the community is necessarily distributed. We don't want the physical location to be a barrier to facilitating translation, collaboration, and partnering. (Interview 6, October 8, 2010)

According to Kahlon, to foster greater collaboration among different kinds of scientists, CTSI created a website that, for all intents and purposes, *is the translational center itself.* I asked her about their process for building the website. She said the website designers wanted to make sure that their site was less a brochure and more an interactive portal. In fact, the website changes in response to the needs of a user: "Especially as we increase partnerships, we need a platform to say what we do. The virtual home had to be presented in the language of the user." An ever-flexible interface that understands the needs of various stakeholders, the system enables key legibility among multiple stakeholders in the translational process. Digital tools facilitate translation between a neurobiologist and, for example, a CTSI consultant and the scientists who escalate the research of their joint interest.

The website consists of functions and capacities one could access and use to complete tasks including networking, finding partners and equipment, and requesting CTSI consultation. There are similarities between CTSI's web-based tools and AstraZeneca's new virtual neuroscience innovation model, iMed. Recall that in 2011, AstraZeneca decided to entirely externalize its neuroscience innovation program, closing all of its neuroscience-focused laboratories and laying off its own scientists and neuroscience staff. AstraZeneca then announced its entirely partner-based approach (see figure 4.5): partnering with university academics virtually. Its new iMed program would not only replace its brick-and-mortar laboratories, but also included tools to manage a global network of academic collaborators.

Creating a virtual center to connect global partners also created (in digital parlance) *engines* (internal and external to AstraZeneca) focused on aligning knowledge in ways that made it discoverable and integratable by AstraZeneca. Virtualization, in terms of its capacity to transcend the barriers of place-based partnerships (see figure 4.6), allows stakeholders to execute the creation or integration of common goals and objectives for knowledge work. In other words, virtual-relationship management and collaboration tools enable a form of knowledge organization that helps manage

## Our Model

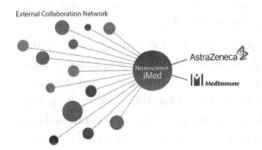

**Figure 4.5**
AstraZeneca's model depicting its new externalization-focused neuroscience innovation approach

**Figure 4.6**
At one of AstraZeneca Neuroscience's two small offices, AstraZeneca stakeholders videoconferenced with partners. (Reprinted with the permission of AstraZeneca.)

immaterialities such as aligning objectives and research interests. Virtualizing the partnering, resource gathering, and consultative processes also has an epistemic function: to render knowledge, stakeholders, and resources legible and connectable.

Kahlon explained that many of the collaborations engineered through the CTSI's website were still social and connective, but the webtool streamlines the kinds of tasks that one might do to turn research into *an enterprise*. The tool categorizes not only the work and expertise within UCSF, but also of the larger UC system. Kahlon described the website's functions and its likely users:

> We don't have the data, but we imagine that those who use it most are faculty——and academic staff. Junior faculty probably really use it, and they are the ones that need to find resources: entrepreneurial assistance, trainees, residents, and fellows. You can get info without signing in, but if you want to request a consultation service, the CTSI consultation service, then you have to sign in. The consultant is paid. (Interview 6, October 8, 2010)

Kahlon paused as if she were going to talk about something sensitive. After reaching for the right wording, she continued, "Charging [internal stakeholders] for university services is a cultural change, and they don't expect to have to pay for university services. Translation is new for all of us. They now have to budget for consulting—in case they want to be connected with the right people to do something *with* their research" (Interview 6, October 8, 2010).

CTSI's internal consultancy, which charged a fee for providing consulting services to academics about how to scale, monetize, or assess their research, is similar to the management consulting model commonly found in the for-profit sector. In fact, the very availability of internal management consultants who provide advisory services for creating a venture clearly speaks of an entrepreneurial strategy at work at CTSI. Yet Kahlon's admission that academics were not accustomed to this model showed just how much the translational shift is clearly toward a model in which academics are being asked to *become* entrepreneurial. In this way, the shift to which she referred was more than a cultural change. It was also about an evolving scientific subject—one prefigured in environments designed to compel entrepreneurial thinking.

With regard to its development, programming, and impact, Kahlon explained that a primary function of the site is also to foster collaboration within UCSF, across campuses, and outside of the university. In particular, the goal for the technology is to foster cross-disciplinary collaboration. "You can search for research cores!" she said excitedly.

*Research cores* are on-the-ground research areas. UCSF had more than 100 cores; most laboratories and researchers were associated with multiple cores. Kahlon said, "A distinct technology that we use in the website is UCSF Profiles. It's a tool, and it's the domain of research networking that is emerging." The CTSI website achieves several functions using the profile tool (see figures 4.7 and 4.8).[6] In addition to providing personalized resources and information to each user, it categorizes expertise areas and groups to which university researchers are tied. Thus, the profile tool categorizes scientists' academic work and provides a knowledge database that could be used internally by UCSF or externally by potential collaborators, investigators, or other stakeholders.

The UCSF Profiles tool uses the language of "expertise mining" to explain its functionality in creating collaborations within UCSF and the larger UC system. The mining excavation metaphor, now thoroughly overused in database language, still captures the evolution of digital search. The UCSF Profiles system is also digitally and informationally linked to similar databases at research institutions nationwide. Thus, while editing one's profile enables personalization, it also feeds an even larger system of categorization, management, and organization. The external functionality of the web-based search tools is interesting. A quick look at the Profiles tool reveals that certain website functions are strictly internal and require

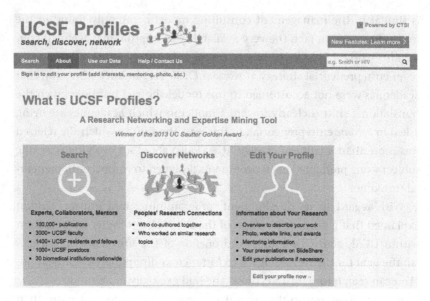

**Figure 4.7**
UCSF Profiles, an "Expertise Mining Tool" created by UCSF and funded by NIH's CTSA program (https://profiles.ucsf.edu/about/AboutUCSFProfiles.aspx)

use authentication. However, many of the search functions are entirely open, inviting internal and external users to mine UCSF's expert knowledge resources.

Kahlon's excitement about research cores was palpable, prompting a search of the website's research cores section. In one area of the site with a cluster of tools presented to the viewer, a small selection of search tools on the right of the screen led to a screen titled "UCSF Cores Search." Enclosed within a rather simplified graphical user interface, a large entry box enabled the user to create searches (see figure 4.9). Most interestingly, the tool revealed how "research core" is understood. Listing 838 resources, 241 categories, and 83 providers, it references a marketplace of tools and capacities, methods, know-how, instruments, and experts all in one place: a perfectly streamlined library of objectification.

Initially, I was drawn to the methodological usefulness of UCSF Cores Search in an analysis of the social construction of science. It provided, if inadvertently, an ingredient list for the objects, expertise, experts, and connections that allow the construction of experimental practices. Thinking

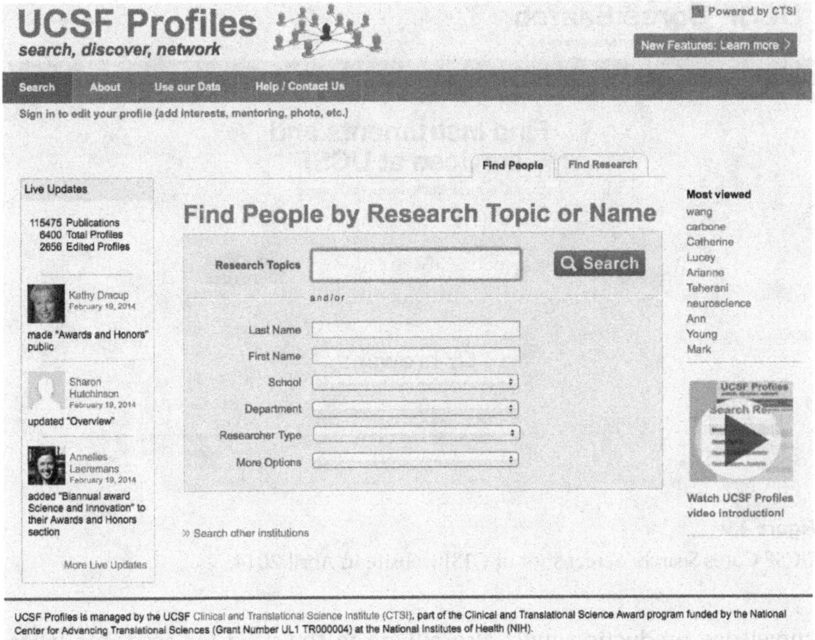

**Figure 4.8**
UCSF Profiles have both open and closed functions, such as requests for CTSI consultancy. The "finding" functions are open and enable external collaborators to more easily find and proposition UCSF researchers. Screenshot of CTSI website in April 2014.

with the amalgamations of actors, objects, and systems accounted for in actor network approaches within the field of science studies, this directory could provide a useful material history of scientific practices. That is, in following one user's use of the resource, one might be able to see this tool as a scientific cookbook—as the tool that helped facilitate critical experimental and research steps. Beyond its implications for scientific practice, the system of categorizations achieved under CTSI's webtools also enables an opportunity to view a set of scientific morphologies.

Post-Fordist models and programs of scientific work (Gieryn 2008) achieved many deterritorializations: laboratory benches became movable and computer monitors flexible and adjustable, data was shared via "the cloud," and knowledge was mythologized as open and organizable (Barbrook and Cameron 1996). In this vein, the very objects and means of

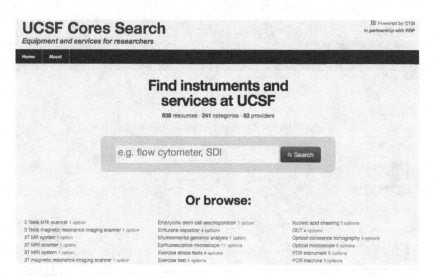

**Figure 4.9**
UCSF Cores Search. Screenshot of CTSI website in April 2014.

knowledge production must also adhere to the need for academic labor to be conceptualizable in terms of contemporary neoliberal flexibility. The formations and formatting achieved through the CTSI webtools are an extension of this reformulation toward flexibility, extended even to include the scientific subject category. Research has come to be understood as resources and assets, capacities and tools, experts and expertise, connections and connectability.

For scientists who use and reference this library of instruments, knowhow, and assets for part of their experimental process, it is a central tool for everyday thinking. Might it inject entrepreneurial considerations at the moment when scientists conceptualize their research—the very contexts where one might use this online tool? It also, therefore, could illustrate the ways that translational science produces real epistemic consequences for the practice and organization of certain kinds of everyday scientific behavior.

Yet another crucial epistemic effect was achieved through these webtools. From an external view, the webtools' "mining" capacities resemble a marketplace in which research interests and questions, scientific histories, publications, research domains, funding histories, and identities are indexed and categorized in service to a *particular* set of logics and users. The system is a digital cart for scientific possibility, allowing researchers in the network to do more than connect to each other. It also engineers a key vehicle to organize assets and resources in ways actually designed for external use. Unlike other webtools, the ability to search UCSF's researchers and expertise is intentionally open to the public.

This raises the question of the mechanics of intimacy in virtual worlds. This question was a concern for the anthropologist Tom Boellstorff, whose ethnography of the virtual world, *Second Life* (2008), explored digital terrains through which new forms of intimacy could be created: intimacies that were not merely reflections or mimicry of the actual world. Boellstorff showed how the mechanics of meaning accrete in unique ways in the decontextualizations of digital environments. Of course, the question remains about whether network systems such as LinkedIn constitute virtual worlds. Nevertheless, UCSF Profiles is more than merely a social network. It also seeks to categorize and organize research areas, expertise, and competencies, and to assess ever-evolving industrial and technological resources. The system's sociality is but one component of its larger function.

Interested in what kind of relationship-building or knowledge-gathering practices occurred through UCSF Profiles and how these practices interacted with or sought to enable intimacies, I called biotechnology investor Michael Hall to ask if he had heard of this online tool. Over coffee in San Francisco's Mission District, he explained that he used UCSF Profiles for several purposes—some of which supplanted older methods of finding information about academic work, of getting "to know" people, and learning about their work in deep ways:

> I've used their system and it's pretty good. It's effective. I have used [it] to try to search for scientists working in the areas that I'm interested in, and I also have contacted researchers directly using the site. It's much easier than having to do all of this networking. It's also better than LinkedIn because I can see exactly what science they are working on. They [scientists using the system] also know that we [venture capitalists] are looking, and so I've seen some scientists start to work more in the directions that we're interested in so that they can be discovered. I like the system. It makes deal flow a lot easier, it also lets me avoid wasting time talking to the wrong people, and it just puts this information at my fingertips. It's a breeze for—actually, I really like their tool, too, because it lets me find exactly who I'm looking for even if I didn't know that I was looking for [them]. It's organized by research and by background, so it helps guide me as to who I might want to talk to. (Interview 30, January 2013)

Two things appear here. First, Hall's use of the tool to learn about potential partners and new research directions shows how CTSI's web-based system helps create and enable knowledge-gathering processes that might otherwise require or rely on intimacy. Specifically, the UCSF Profiles tool helps categorize scientists in relation to their existing projects. Such categorization may only become visible after extensive social networking, but it works in tandem with the actual interactions of university and industrial partners. It not only possesses part of the self-presentation capacity enabled on professional networking sites such as LinkedIn, but is also a resource developed for partners with the goal of specific kinds of connectivity in mind.

However, Hall's contention during our conversation also confirms a second point. A large part of these mining tools is presenting knowledge projects in ways digestible for venture capitalists, and therefore the language eases barriers between university research work and partnership with venture capitalists and the biopharmaceutical industry. Such software tools facilitate and even guide scientists to the kinds of scientific questions that may be relevant for potential external partners like Hall. His use of Profiles suggests that the CTSI system does more than merely provide results. Hall's searches led him to the work of researchers of whom he had been unaware, suggesting that the CTSI discovery system also enables a crucial "filling in" of scientific gaps for potential partners. In this way, the tool becomes an agent of translational thinking. This implication may also be relevant for those on the other side of the search.

Beyond that, the CTSI website constitutes a kind of extensional formatting attached to what one might think of as everyday scientific practices— that is, the way a scientist might engage in a set of laboratory experiments, which historically may have had no life outside of the laboratory. A discovery engine in which all of a scientist's work is locatable and conceivable in terms of translatability and application and that enables that work to be executable, findable, and "leverageable" as part of a commercial aim helps allow scientific activity and resources to be reimagined as an asset. It is thus unsurprising that soon after the CTSI website launched, an additional stage in the assetization (Birch 2017c) of science occurred in 2008 when the United Nations Statistical Commission decided that R&D spending should no longer be treated as expenditure within the Systems of National Accounts (SNA), but was to be recategorized: "Following the SNA 2008 decision, then, R&D spending would be reframed as an investment because it produced an asset, in this case *scientific knowledge"* (Birch 2017c, 1691, original emphasis).

Thinking with the mechanics through which scientific knowledge gets reframed in relation to the market, one can see that the role of *connections* in the translational endeavor—connecting biological material to disorders, biological material to discourses, markets to new therapeutics—extends to the institutionalization of translational science in relation to its effects on scientific thinking. Webtools can be seen as vehicles to change how research and researchers are conceptualized in relation to a capacity for commercialization, connection, investment, and entrepreneurship. Here translational research and science become a means of *epistemic action.*

By epistemic action, I refer to actions that "are not used to implement a plan, or implement a reaction; they are used to change the world in order to simplify the problem-solving task ... actions that make mental computation easier, faster, or more reliable" (Kirsh and Maglio 1994, 513–514). My fieldwork, observations, and analyses impelled me toward the proposition that the whole of translational science and research is positioned to compel epistemic action. That the CTSI website reorganized research, expertise, and researchers to be easily managed using its database and search function thereby changed the world to enable indexability and search functionality.

In this meeting of epistemology and material tools (such as the CTSI website), we can derive guidance from the history of cognitive anthropology and especially its reflection on the use of the material and social worlds in thinking practices. One concept that targets this interplay is the "cognitive artifact," which Roy D'Andrade (1995, 210), invoking the work of Edwin Hutchins, referred to as "a physical structure that helps humans perform some cognitive operation. Cognitive artifacts range from complex machines such as computers to simple devices such [as] paper and pencil." Thus, beyond philosophically inflected concepts such as epistemic action, there is a need to consider the roles accorded to tools and artifacts in social processes. While the CTSI webtools consisted of actual tools (i.e., web search and profile and account creation tools), I also conceive of the CTSI website as *an environment.* As such, it is important to understand how the website's constituent tools produce cohesive sets of practices. The tools focus on data organization, entry, mining—all speaking to a common system with articulated goals. CTSI as a cognitive artifact brings up a critical question of the ends to which CTSI compels thinking and practice.

Catalyzed in part by $112 million in funding from NCATS in 2011, the CTSI webtools provide an example of how translational science's institutionalization at the university creates entire systems. The UCSF portal is not only a means of connection; it also includes transformations that enable potentially all research and research capacities to be organized according to one's needs or agendas, or more importantly to the needs and agendas of others. To the extent to which the tool compels one to *think* of scientific practice in particular ways, a question emerges about how the institutional work of translational science—especially work focused on finding and aligning interests or creating epistemic legibility—is facilitated by creating systems.

Creating software programs to identify knowledge synergies and catapult researchers into the known paths of venture capitalists represents a significant moment in how universities handle innovations and treat university knowledge. In fact, this mode of technology-assisted, internal-discovery processes and internal consultancies reflects a real shift on the part of universities toward becoming knowledge-management organizations focused on the "efficient" organization, management, and monetization of knowledge resources and assets (Mirowski 2011; Shore and Wright 1999; Strathern 2000).

The importance of these website technologies (known as *information engines*) produces an interesting picture juxtaposed against the Booz & Company report that discussed means to create external innovation systems. (As an aside, the facilitative nature of the website also shows the need for greater anthropological and ethnographic investigations into websites, since they are becoming the *means of things* rather than just sources of flat information.) The report similarly underscored the importance of digital tools in scientific practices—an area that has only gradually attracted scholarly attention (Beaulieu 2002). Nevertheless, the UCSF center's website is an integral system through which exchanges of capital, expertise, networks, and access are mediated and tracked. To underestimate the power of this information technology in the system of translational science would be to overlook how the commercialization of science encourages efficient management of university assets. Here, management is the primary bureaucratic force through which systems were produced and university work and resources sustainably organized.

The implications of both the CTSI technology and the management practices around it (such as UCSF's consulting service) show the durable work of systems and the role of systems making as part of the translational paradigm. These point toward an analysis of the role of environments, technologies, and materialities in the creation of epistemic effects.

## 4.5   Technology, Design, and Contoured Actions

To more clearly delineate the way that translational systems "seek" to durably shape actions around knowledge, it is important to map how such systems are *designed* to shape knowledge actions. How is design implicated as part of translational visions? To what extent are translational actions—scientific envisioning, partnering, investing, knowledge seeking—the product of systems designed to compel "translation"?

I learned more about the role of design in the functioning of the CTSI web system through conversations with Kahlon. According to her, the CTSI web system went far beyond merely enabling the user to search. Instead, the system compelled action among its users—whether scientists, university administrators, or prospective investors. She stated, "The [webtool's] everyday search and discovery functions are fine [i.e., sufficient] for older researchers, but these technologies are trying to put together information and get people to act upon it. There is a side to networking beyond mere discovery." In this way, I began to think about this tool as key to the mobility of knowledge necessary under TN. Kahlon's elucidation answered the question of how scientific resources (human or otherwise) became actionable for investors, potential collaborators, or partners. Yet questions about knowledge mobility call for a question about the technology itself. How was the CTSI digital environment designed to "get people to act upon" the information on the CTSI website? In thinking about the website's actual "architecture," I began to reflect on the coding and programming that lay underneath the design (see figure 4.10)—a system programmed to send users from one entry point to the next, which automatically recognized certain search terms and replaced them with others, and that transformed raw data into usable forms for the site user.

To think of these webtools' backend design would also be to conceptualize the many pathways, predetermined goals, and literal "logics" encoded in the system and designed to not only assist one's thinking (i.e., in the way

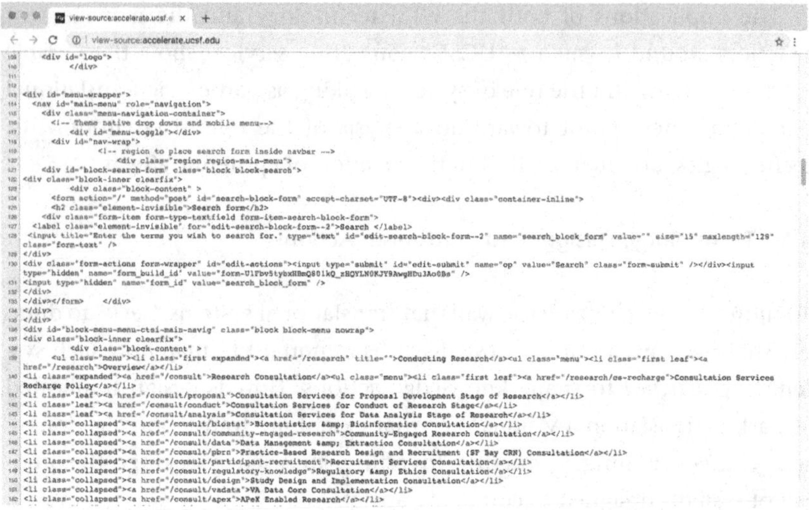


```
    <div id="logo">
          </div>


<div id="menu-wrapper">
  <nav id="main-menu" role="navigation">
    <div class="menu-navigation-container">
      <!-- Theme native drop downs and mobile menu-->
          <div id="menu-toggle"></div>
      <div id="nav-wrap">
                <!-- region to place search form inside navbar -->
                    <div class="region region-main-menu">
    <div id="block-search-form" class="block block-search">
<div class="block-inner clearfix">
          <div class="block-content" >
      <form action="/" method="post" id="search-block-form" accept-charset="UTF-8"><div><div class="container-inline">
      <h2 class="element-invisible">Search form</h2>
    <div class="form-item form-type-textfield form-item-search-block-form">
  <label class="element-invisible" for="edit-search-block-form--2">Search </label>
 <input title="Enter the terms you wish to search for." type="text" id="edit-search-block-form--2" name="search_block_form" value="" size="15" maxlength="128" class="form-text" />
</div>
<div class="form-actions form-wrapper" id="edit-actions"><input type="submit" id="edit-submit" name="op" value="Search" class="form-submit" /></div><input type="hidden" name="form_build_id" value="form-UlFbv5tybxHBmQ801kQ_zHQYLN0KJY9AwgHDu3Ao0Bs" />
<input type="hidden" name="form_id" value="search_block_form" />
</div>
</div></form>    </div>
</div>
</div>
<div id="block-menu-menu-ctsi-main-navig" class="block block-menu nowrap">
<div class="block-inner clearfix">
          <div class="block-content" >
      <ul class="menu"><li class="first expanded"><a href="/research" title="">Conducting Research</a><ul class="menu"><li class="first leaf"><a href="/research">Overview</a></li>
<li class="expanded"><a href="/consult">Research Consultation</a><ul class="menu"><li class="first leaf"><a href="/research/cs-recharge">Consultation Services Recharge Policy</a></li>
<li class="leaf"><a href="/consult/proposal">Consultation Services for Proposal Development Stage of Research</a></li>
<li class="leaf"><a href="/consult/conduct">Consultation Services for Conduct of Research Stage</a></li>
<li class="leaf"><a href="/consult/analysis">Consultation Services for Data Analysis Stage of Research</a></li>
<li class="collapsed"><a href="/consult/biostat">Biostatistics &amp; Bioinformatics Consultation</a></li>
<li class="collapsed"><a href="/consult/community-engaged-research">Community-Engaged Research Consultation</a></li>
<li class="collapsed"><a href="/consult/data">Data Management &amp; Extraction Consultation</a></li>
<li class="collapsed"><a href="/consult/pbrn">Practice-Based Research Design and Recruitment (SF Bay CRN) Consultation</a></li>
<li class="collapsed"><a href="/consult/participant-recruitment">Recruitment Services Consultation</a></li>
<li class="collapsed"><a href="/consult/regulatory-knowledge">Regulatory &amp; Ethics Consultation</a></li>
<li class="collapsed"><a href="/consult/design">Study Design and Implementation Consultation</a></li>
<li class="collapsed"><a href="/consult/vadata">VA Data Core Consultation</a></li>
<li class="collapsed"><a href="/consult/apex">APeX Enabled Research</a></li>
```

**Figure 4.10**
Coding behind UCSF's CTSI web interface (https://profiles.ucsf.edu/about/AboutUCSFProfiles.aspx)

that autosuggesting mechanisms in search tools help suggest proper spelling or related searches), but also compel the user toward specific *actions*. In this sense, the picture of the programming code underlying the CTSI website began to look like a kind of ideology given its design directed at steering users toward translational actions informed by a set of overarching ideals about the aims of knowledge. At the same time, the CTSI website became the birthplace for a kind of epistemic subjectivity to the extent that it could contour knowledge-related actions and outcomes for users. These actions that the CTSI site was designed to elicit were not arbitrary. Instead, the design reflected goals elucidated within CTSI's larger mission, which itself reflected a changing and entrepreneurial landscape of academic laboratory science and that, as argued earlier, was inextricably implicated in financial shifts in the pharmaceutical sector. Hence, an ecological conceptualization (Star 1995) of the many built environments that enabled TN (such as the CTSI website) helps us to think about processes, materialities, and designs through which TN compels actions.

## 4.6 Environments

As part of my investigation into how translational research worked on the ground, I sought access to TN laboratories at a university. After months of networking, I was finally able to secure an invitation to observe a laboratory at a Northern California university. However, my contacts at this particular laboratory were hesitant to share much of their work with me—especially as they were embarking on work that had particular intellectual property risks. In keeping with translational science's rhetoric, they articulated strong concerns about the risks of sharing potentially "groundbreaking discoveries." The laboratory's head was concerned about my sharing (even inadvertently) details about their actual research. I had not anticipated this barrier, and yet it made complete sense in the context of the rapid commercialization of university research. Their hesitation contrasts with the conditions at work during Paul Rabinow's (1997) early ethnography of a biotechnology company's development of the polymerase chain reaction in the 1990s. In that ethnography, one notices that his interactions were much more open than my interactions among those within this particular laboratory. Therefore, as requested, I have anonymized the participant names, university, and specific laboratory center within which I worked, as I have done with many individuals throughout this book.

As I navigated the campus for the first time, I prepared mentally for the social interactions that I would likely encounter and tried to remember the laboratory hierarchy. As an anthropologist who worked at the intersections of neuroscience, translational medicine, and patient communities, my initial focus was on the social worlds within which the laboratories existed and how they were impacted by the changing landscapes of translational science and research. My ethnographic assumptions led me to think about people and their relations.

Yet the building was a surprisingly crucial starting place for this rather complicated puzzle. The building design, boundaries, and spaces became critical material artifacts. This university had recently erected a spate of imposing, modern glassy structures to house an amalgam of centers and laboratories dedicated to translational research and medicine. Clustered in one area, this new "health campus" was to be the space of "world-class innovations," according to one dean. As I explored, I began to reflect on how these structures might affect what stakeholders thought about health and how their conceptualization might affect the outcomes of subsequent work.

It was here that I began to think about the relationship between the design of translational buildings and that of webtools, as they were both designed as environments. In both cases, the creation of novel environments and infrastructures (digital and built) was thought integral to the realization of translational goals on the ground. In fact, one of the things that demarcates translational research from prior efforts at application-oriented research paradigms has been the creation of commercial, digital, and built environments and infrastructures.

## 4.7  Listening to Architectures

When I first got to know the building that housed the laboratory, I was unexpectedly captivated by its design. The building stood out against more dated buildings across campus. A plethora of angles, it was a study in glass and steel. From the outside, one could clearly see into workspaces and hallways, and the views into the building were impressionistic. Details were not clearly discernible; people, objects, and light came into view as impressions. The external architecture caused one to imagine the inside as light and airy, clear and open. The olive and blue planes of glass that clad the building's outside dominated the view. The building was a material force itself.

Although there were several entrances, a voluminous, multistory foyer clearly marked the primary entry. Inside the foyer, my eyes were drawn to the extensive vertical space. Colors were neutral and soothing; furniture in the public spaces was modern and low-slung. I later found renderings of the interiors and exteriors of the building before its construction. Even in the digital renderings, which displayed nothing more than an architectural imagination, the foyer's waterfall of light and its dramatic ceiling— the features that first attract the public's attention—were reminiscent of a cathedral. Decidedly modern, the building's interiors conveyed a sense of grand lucidity.

Another element of the design caught my attention. Throughout the building, one found what seemed to be endless expanses of glass. Inside the building, the interior walls, including hallways and the conference-room walls, were glass. Some hallways were set on a curve. The multiplicity of gentle curves around an open atrium created the feeling of being inside a translucent wave. The building seemed to be a flowing collection of laboratories and centers dedicated to advancing the larger project of health.

## 4.8 The "Alzheimer's Laboratory"

I received an invitation from prominent neuroscientist and prolific researcher Tom Carlisle to spend time at one of the scientific labs of the larger Alzheimer's laboratory. Carlisle did not spend as much time in his laboratory as he did in his other office across campus. However, as the principal investigator for several grants that keep the laboratory going, his influence was constant. Very reticent to talk about the laboratory's hypothesis regarding neuronal death in relation to Alzheimer's disease—a reticence that may have connected to his desire to create a diagnostic based on the laboratory's work—Carlisle was enthusiastic in talking about the new laboratory space.

During the tour he provided of the whole building and his laboratory, he talked about the trend toward open knowledge, sharing, collective work, and decreasing knowledge silos in science. He used the example of a recent move by an academic publisher to prevent open-access research. For him, being "closed" was antithetical to the spirit of science. Perhaps reflective of a Northern California ethos, Carlisle's speech was peppered with frequent uses of the words *new* and *future*. I followed up with him about his own collaboration efforts. He reported frequent collaboration with other members of campus institutes. I then asked if he thought the laboratory arrangement had anything to do with this collaboration, and he responded, "Probably not." Tangible instruments such as funding stipulations tended to facilitate more cross-disciplinary collaboration, he said.

Relatively new, the laboratory was located on the top floor of a glassy building. According to Carlisle, the lab focused on understanding the neural mechanisms that coincided with Alzheimer's, to determine a clearer (or less fuzzy) set of clinical markers for the disease. The goal was to create a better set of diagnostic criteria for early detection, and accordingly, early intervention. He noted that the challenge with Alzheimer's, "as is the challenge with many degenerative disease categories, is demarcating it from typical shifts in memory that occur with age." Thus, the work of this laboratory was one of isolation, characterizing "normal aging" in the brain and isolating the unique features that might more clearly characterize Alzheimer's.

Anthropologists (Kaufman 2006; Good, Subandi, and Good 2007; Graham 2006; Hinton et al. 2006; Whitehouse, Maurer, and Ballenger 2000) and work by Lock (2013) have drawn attention to the various constructions of Alzheimer's and dementia by clinicians and practitioners. Even outside neuroscience, the Alzheimer's category has challenged clinicians, and the problem was not lost on this group of researchers. The incredible diagnostic and pathological difficulty of Alzheimer's has impaled the field and grabbed headlines (Belluck 2011). Nevertheless, the ostensibly mundane laboratory science of working with cells and proteins, and creating experiments using animal models—when conceptually tied to the sociomedical fact of Alzheimer's—enabled a particular kind of value under the banner of translation.

The laboratory I visited is prolific in terms of deliverables: research articles, citations, and research funding dollars procured. Headed by Carlisle, a prominent neurologist and professor, it embodies the kind of work that the larger university touts. The laboratory is also symbolic of future health innovations. It constitutes a crucial ethnographic space on its own: low-level experiments are done in the name of Alzheimer's, a disease that is pathologically complex but that also indexes larger questions about the brain, the aging body, and the nature of clinical demarcation. Alzheimer's is also an emotional disease. Sara, a Chinese lab worker, explained that this project resonated with her:

> Back in Shanghai I have my grandmother. … She has Alzheimer's. It's been very hard to watch. I love my grandmother and seeing her decline has been very difficult. My mother and some other family take care of her. It makes me feel good to be working on Alzheimer's. It makes me feel like all of this work that I do is for something good. I don't get back to Shanghai very often but I always make a point to visit her. (Interview 25, December 5, 2011)

The language that lab workers used to describe their research clearly reflects the connection of this laboratory work to the broader aims of health and specific disease categories, even when it was more about specific, yet otherwise typical, biological processes (degeneration being a typical biological process). I had often thought that the "translation" lab workers made of their microscale work in disease categories was created simply to accommodate me as an outsider. I thought perhaps they were doing this to make these small microexperiments make sense in a larger view. Those with whom I spoke, however, similarly narrated their own experience as working on a disease even after I was no longer a newcomer. Tiffany, another worker in the laboratory, often found popular news articles about Alzheimer's and would post them in public view for other lab workers. During laboratory meetings, there would be informal conversations about the news articles or pieces of popular media about Alzheimer's disease.

## 4.9   Goals "in Mind"

Miku worked on the molecular mechanisms of neuronal death—a primary Alzheimer's marker. Tiffany, who once made a macabre joke that theirs was the "death lab," tested degeneration using rat models. According to Carlisle, many therapeutic opportunities could be tied to this research. Miku seemed interested in this possibility of application, whereas Tiffany rarely articulated her everyday lab work in therapeutic or otherwise grand terms. I had asked Miku about what possibilities this research could create in terms of therapeutics. "Well," she said tepidly and with a tone of uncertainty, "if you get to understand this stuff more intensely, then maybe you could try to create prevention strategies." I asked her what she meant by strategies. "I mean pills. I guess maybe there could be other prevention methods, but pills are the main way that this stuff works." I later asked Miku if she recalled thinking about pills as she began to set up experiments and propose new research priorities. She confirmed that she did definitely "think about the end product" as she did this kind of work.

I was also interested in the specificity of their lab work compared to basic neuroscience laboratory research because neuronal death is a characteristic of many neurological diseases and the problem of isolating Alzheimer's meant this research was actually rather broad. When I asked Miku about the difficulty of balancing what looks like basic research with the expectations of specific applications, she quickly nodded in agreement but did not respond further. During the time that I knew Miku, she tended to keep laboratory life within the confines of the lab. An otherwise super jovial person, she tended to do her work, go out for lunch, finish her work, and quickly leave for home. She did not tend toward long, tortuous reflections on her work or that of the larger laboratory, at least in the context of our conversations.

Carlisle, however, was able to provide a more specific, but limited, response to my questions. The work of this laboratory and its connection to the translational project was about focus: "We're doing this work with a goal in mind—the discovery of novel ways to make a difference in this disease." Carlisle's punlike phrase, "goal in mind," coalesced with my notion that TN comprises a particular way of thinking—a mode that may constitute one important differentiation between TN and basic science. From both Miku's contention about "finding pills" and Carlisle's response about "goals in mind," one sees the articulation of what I call *translational thinking*. While I discuss this briefly in the latter portions of this book, I use this term to refer to the effect of prior presumptions of a psychopharmacological, pharmaceutical, or device-based product that becomes both the pretext from which scientific designs are drawn and the measure for how scientific successes are assessed. A technological heuristic, it produces upstream effects, determining which experimental questions are asked, which data disregarded, and which solutions envisioned. Translational thinking not only constructs the very objects of investigation, it also constructs its subjects.

Additionally, Carlisle's suggestion about "making a difference in this [Alzheimer's] disease" relied on an understanding of translatability between the work inside this laboratory using animal models (and disease models specific to these animal models) and "Alzheimer's disease." This distinction about *goals* as key differentiating factors between basic and translational research corresponds with early laboratory studies (Knorr-Cetina 1981; Star 1989) about the inherent everydayness of laboratory work. The external (social) and internal (scientific) significance accorded to laboratory work is a key component in the production of what the work in the laboratory means. Bruno Latour's (1983, 146) discussion about the displacing nature of translation targets the way that laboratory objects can stand in for real-world objects: "Like all translations, there is a real displacement through the various *versions*." Keeping these versions together, according to Latour, is the principal means by which laboratories maintain the power to speak to macroscale problems using microscale laboratory work,[7] which is the essence of bench-to-bedside translation.

The movement of the fact from the laboratory to the world of disease categories is at the heart of translation for Latour. Heretofore, scholars have had to theorize about how facts travel, since few pathways were clear.[8] Thus, it is essential to understand the role of translational science as representing a new point in the history of these investigations: a useful new configuration in which boundaries between the outside and inside are rendered open, and this openness rendered explicit. Translational science relies on this categorization of valuable and useful work, as well as the identification, binding, and codification of relevant research options. Translational science therefore not only includes a moral orientation toward bringing knowledge out of laboratories (this is not just biological research, it is "*Alzheimer's* research")[9] but also includes infrastructures, legal accommodations, private partners, state institutions, and discourses to help support the process.

In the piece "Give Me a Laboratory and I Will Raise the World," Latour (1983) sought to show the means by which activities inside the laboratory were understood to be exportable to the larger problems in the world. For him, "Something is happening in these [petri] dishes that seems directly essential to the projects of these many groups expressing their concerns in the journals." In the latter part of the article, Latour (1983, 155) uses Louis Pasteur as an example to think with and offers a useful proposal for considering innovation and its presuppositions:

> That this metaphorical drift, which is made of a succession of displacements and changes of scale ... is the source of all innovations is well known ... (For our purposes here, it is enough to say that each translation from one position to the next is seen by the captured actors to be a faithful translation and not a betrayal, a deformation or something absurd. For instance, the disease in a Petri dish, no matter how far away from the farm situation, is seen as a faithful translation, indeed the interpretation of anthrax disease. ... It is useless trying to decide if these two settings are really equivalent—they are not since Paris is not a Petri dish—but they are deemed equivalent by those who insist that if Pasteur solves his microscale problems the secondary macroscale problem will be solved.

This model suggests that there is a work of *faith* in the rectification, in bringing together the microscale and the macroscale. This was certainly the case for my interlocutors at the Alzheimer's lab, for whom their microscale research about the workings of neurons (a very common, mundane process) was easily mapped onto the larger disease category and the possibility of discovering an intervention. Yet the very fidelity of these meanings, which bound the petri dish in Pasteur's lab to the social category of the disease to which it was attached, could only occur through the key tool of discourse. Specifically, the discursive mechanics involved in maintaining meanings is a key process in the work of translation. Thus, in the example of the Alzheimer's lab, even before Carlisle hired these laboratory workers—before Tiffany and Miku were doing their work—this was already an "Alzheimer's lab."

This is not to say that translational science is reducible to nothing more than discourse. Rather, this points to the importance of the discursive work in translational science, to codify it as work connected to larger disease states and goals. Although displacement is especially useful in thinking about translational science, where scientific claims move from the "petri dish" to outside, this emphasis ignores the work of discourse that ties the scientific claim to other translational efforts (medical, technical, and epidemiological) and allows various kinds of research to be constituted as health. It is, importantly, in the discursive attachments where TN becomes understood as meaningful in terms of health. Here, I refer both to social discourses (e.g., "this health campus is the future of health") and internalized discourses (e.g., "we're working in Alzheimer's").

In Latour's (1983, 164) example, aside from seeing the work of displacement, it is necessary also to notice the way that a given *environment* (the lab) and its associated epistemologies engineered and reproduced important discourses about what is done within the laboratory:

> Thanks to a chain of displacements—both of the laboratory and of the objects—the scale of what people want to talk about is modified so as to reach this best of all possible scales. ... Then everything they have to talk about is not only visible, but also readable, and can easily be pointed at by a few people who by doing this dominate.

However, focusing merely on how discourse facilitates such displacements—or in a more limited sense, the "contract" of translation—obscures the importance of material environments for contouring discourses and subjectivity. It is essential to examine how translation narratives give life via material processes in the laboratory and to assess the effects of these environments on scientific practices.

## 4.10   An Architecture for Fantasies

The Alzheimer's lab was part of a larger cluster of centers and institutes that had been physically located near one another. The university administration's push toward a modern, biomedical health sciences center was itself reflective of a larger imagination about health. Even the visions regarding the new campus reflect a narrative about the ability to "build" an ostensibly complete health solution. In other words, biology labs were established adjacent to policy research offices; their proximity to a new research center focused on genetics was also no accident. The idea was that this interdisciplinarity would be the means by which large problems in health could be solved. These dramatic visions about the future of health were reflected publicly. While in Northern California, I visited multiple sites where buildings were being erected and programs founded. I attended one particular groundbreaking for such a project at the University of California. The center's director offered a rather biopolitical context for the new building's role in emergent health innovation:

> We are indeed at the beginning of a new era of scientific research and innovation that promises major advances in the biological sciences that will profoundly impact human health. ... Knowledge of biology and how the body works is reaching a level of sophistication that could change the paradigm of health care. By treating and preventing diseases at the molecular level, a new generation of doctors, clinicians, and researchers will redefine how disease, aging, and life itself is understood and experienced. (Robert Tijan, faculty director of the California Institute for Regenerative Medicine, at ribbon cutting and dedication of new Biomedical and Health Sciences Center; quoted in Rodríguez 2011)

In this narrative, one finds the kinds of displacements in the Alzheimer's lab in which everyday laboratory work is connected to the larger category that is Alzheimer's disease. Thus, to illustrate the importance of looking at the designing of the Alzheimer's laboratory as a specific environment, I demonstrate that much translational science is informed and pushed by *desires* for a seamless shift from "bench" to "bedside." Although it may seem obvious, designs that hide the many fissures and challenges in the translational pathway obscure much of the complexity of the process of "translating" a basic discovery into a health solution. Creating translational infrastructures is powered by octane-fueled envisioning. One article about the ethics of translational science programs provides a great example of how embedded desire is in the animations around TSM: "The current 'celebrity status' of translational research reflects the aspiration to facilitate the progression from basic discoveries to new interventions, and from their approval to their widespread use" (Eyal and Sofaer 2010, my emphasis). Accordingly, the NIH went through an extensive research phase in which their stakeholders spoke with scientists, clinicians, universities, and private industry to try to rethink infrastructures according to a grand vision for health. The imagined *legibility of facts* (e.g., that a fact about cellular degeneration patterns in a lab rat could be transformed into a clinical diagnostic test), and imaginations about the ease of transferring knowledge from one domain (a department or discipline like neurobiology) to another (e.g., engineers creating testing tools), also heavily relied on vision. This led to an interesting commonality that tied together scientists, pharmaceutical companies, and institutions as part of an imaginary.

Even in the many graphs, diagrams, and flowcharts that outline the processes by which translation ought to occur, one finds digitally produced images in which various disciplines and aprpoaches are perfectly separated, separable, integratable, and then connected by arrows to gesture toward translation. To be clear, I do not aim to paint a critique of translational programs in which they suffer from delusions of ease. My ethnographic experiences from the Neurotechnology Investing and Partnering Conference and my conversations with neurologists provided clear evidence of stakeholder concerns regarding the ineluctable difficulty of translation. Instead, I point to *the essential work of imaginaries* in the elucidation of translational programs and translational work—imaginaries made clear through an examination of the design of the laboratory environment itself.[10]

### 4.11   Glass Labs and Ghost Corridors: The Material Life of Transparency

To the casual observer, the Alzheimer's lab seems encased in a glass box (see figure 4.11). Attached to an explicitly named TN center, the laboratory's glass walls and internal walls blend. Space separation is difficult to discern because the walls separating labs and offices are all glass. The sense of transparency is immediately disorienting for a new person. At least once a month, I would make the same mistake and walk into an adjacent lab because I was disoriented. Miku hated these glass walls. She once said to me, "I always feel on the spot." As with most of the workers in her laboratory, much of her day was spent behind these walls working—in her case, with lab mice. Occasionally, she and I would have a lunch of Vietnamese fresh spring rolls from a nearby restaurant. We often talked about how to induce neuronal degeneration. However, when we were not talking about that, we frequently spoke about the glass walls. "I feel like ... I feel so *exposed*," she said. I felt similarly inside the space.

**Figure 4.11**
An image from an architecture firm that designed the Boston-based translational laboratory, Broad Institute, which is quite similar to the one in which I worked in Northern California. A prominent architectural firm that specializes in modern laboratory design also used this image to illustrate the designs for transparency. *Source:* Perkins + Will.

At one side of the Alzheimer's lab was a hallway. One could see through the glass hallways all the way into the building spaces beyond. I later learned that the collocation of multiple, interconnected laboratories operates through the use of what in architecture are termed *ghost corridors* (see figure 4.12), a provocative term for an essentially invisible hallway that spans the length of multiple laboratories. The amalgamation of laboratories and the extensive use of glass walls shaped Miku's anxiety. I asked her about her initial statement on feeling "exposed." In her elucidation, she talked about feeling as if she were constantly observed. I found her statement intriguing: narratives of openness and transparency attached to open designs usually invoke a rhetoric of freedom, yet glass walls and open spaces can also simultaneously become means of surveillance (see figure 4.13).

One day, Miku put up small curtain covering a portion of the glass wall adjacent to where she worked. It became a minor source of jovial conversation inside the lab. I later found that her decision to put up curtains was not

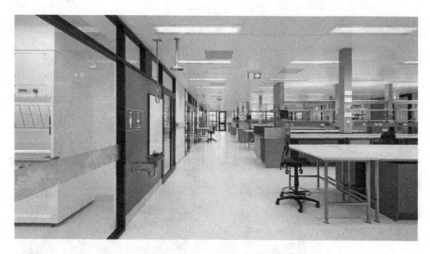

**Figure 4.12**
Laboratory "ghost corridor" circulation. Copyright © Christopher Frederick Jones.

**Figure 4.13**
Stanford University's Clark Center open lab. *Source:* Perkins + Will website.

dissimilar to actions by other scientists working in overly transparent, glassy laboratories. In fact, there was a similar occurrence by a scientist working in a newly redesigned, open lab at Princeton University.[11] Yet there is also a crystalline irony here. Miku's research on neuronal death involved studying laboratory mice that lived their entire lives inside clear Lucite boxes on lab tables. There were no illusions regarding the need for those Lucite boxes to be amenable to laboratory observation. Nevertheless, there was something provocative about this parallel between the Lucite boxes housing the lab mice and Miku's "Lucite laboratory." In one sense, the glass-bounded lab itself could be seen as part of an experiment, which like the Lucite box was designed or redesigned toward particular experimental knowledge goals.

Miku was not the first person I had met at the laboratory. Sara and Tiffany had both been very friendly during our initial conversations. But Miku and I had developed something of an impromptu relationship: we were the same age and could talk comfortably. She became my principal interlocutor in the laboratory. A relatively recent transplant to the laboratory and the region, she had come to California to conduct laboratory research as a PhD student. Like many of the laboratory's mostly international lab workers, Miku was married. Unlike the other lab workers, she also had children. Her husband, Justin, was a postdoctoral fellow in biology also at the university. He worked at a laboratory not far away. Occasionally, Justin would come up to Miku's floor, walk to the laboratory, and wave through the glass, inviting her to lunch. According to Justin, the campus had relocated several labs in order to segregate a portion of the campus devoted to science and engineering, relegating it to a new science and engineering cluster in a newly developed part of campus. "The move was awful," Justin said, explaining the logistical difficulties involved. "We had movers come, and several professors didn't want the relocation, but it was part of the strategic plan." The need for greater innovation via translational science was part of a strategic plan in which the university's new health science and science campuses were centerpieces. I began to think of the relocations to which Justin referred and the specific design of Miku's laboratory and building as part of a connected discourse around "innovation."

To achieve a greater understanding of the design of the "transparent laboratory" in which Miku worked, I spoke with John Miu, the principal architect of the Center for Health and Biomedical Innovation that housed Miku's laboratory. Miu was a distinguished, gray-haired man with distinctive mauve glasses. An esteemed architect, he had designed a myriad of academic science buildings; his eponymous firm practiced what he called *architectural minimalism*. Inside his office, one felt the opposite of minimalism; a bevy of awards, furniture, and models littered his massive space, a rehabilitated warehouse in "SoMa," San Francisco's converted (So)uth of (Ma)rket industrial area, now home to young tech companies and a burgeoning nightlife. As at many such firms, several dozen young architects and interns do much of the work of designing, planning, and managing client projects to facilitate Miu's building projects. As I entered his office, his space was abuzz with fashionable young architectural designers, engineers, and project managers. Grateful for the opportunity to meet with Miu, however briefly, I asked him about his thoughts and intentions with respect to using so much glass in the science building. I wanted to know how this functioned in terms of the larger rationale for the building's design.

"Glass walls inspire cross-disciplinary conversation, collaboration, and innovation," he said. Spry and energetic, Miu popped up from his Eames-style office chair to walk to the other end of his large office. I followed him around his office as he showed me pictures of interior designs for Silicon Valley technology companies, which he then used to talk about the design of the university building:

> See the open space and attention to common spaces? That's how you create an innovative company. Closed offices and closed-off rooms are a thing of the past. Knowledge has now got to be mobile and openness breeds open thinking, and creativity. Think about this [design for the center] as representing the breaking down of walls and barriers. Knowledge for the new economy. (Interview 27, January 2012)

Miu's thoughts about how to compel "open thinking" brought together the materiality of the laboratory and a set of logics regarding knowledge. In this way, designing environments of knowledge became a means of executing neoliberal fantasies about innovation, information, communication, collaboration, and an emerging post-Fordist modernity. For Miu and others, glass walls inspired and compelled subjects to become certain kinds of knowledge workers. Toward the end of our conversation, he talked about how glass walls spoke to modernism in the way that glass erected boundaries but also removed them. After our discussion, I returned to the glass building. In the foyer, light would often move in from skylights, through the atrium, and spread throughout the interior hallways and corridors as the day passed, which only further cemented my sense of the building as a kind of elaborate Lucite box.

The idea that designing for free flows of knowledge might compel interdisciplinary collaboration served to imprint both social and epistemic goals via lab design. In this way, Miu's statement that "glass walls inspire cross-disciplinary conversation, collaboration, and innovation" included a view of design as a directive or prescription in which the goals are not merely social, but also imbibe ways of thinking ultimately affecting the kind of knowledge created, a mode that presumes that mere interaction or communication will transform the kind of knowledge that emerges. Of course, open laboratory design is nothing new. Thomas Gieryn's (2008, 796) work traces the evolution in science laboratory design and its connections to market "insistences of late capitalism."

The rearrangements that compelled the lab design in which Miku works and the campus rearrangements of which Justin was part when his laboratory was relocated reflect the displacements justifiable under TN narratives. They also became codified as environments necessary to achieve the visions of science and medicine under translational science. Interestingly, I found traces of these same discourses about the material conditions for scientific innovation in architectural discussions about the future of laboratory design.[12] In an article titled "Trends in Lab Design," the author states, "Science functions best when it is supported by architecture that facilitates both structured and informal interaction, flexible use of space, and sharing of resources" (Watch 2010).

Sociality, flexibility, spatiality, and transdisciplinary collaboration were all seen as the ideal in a modern, global, and ostensibly open and transparent knowledge economy. Thinking about these desire-laden discourses attached to the narratives regarding the design of open laboratories and "social buildings," I wanted to return to the laboratory itself. Miku's experiences of "openness" betrayed the optimistic visions of "collaboration" and "flexibility" implicit in such discourses about the new, modern laboratory. Nevertheless, I remained interested in the *environments being produced* under the translational imaginary. The rearrangement of university workers, the narratives attached to the design of the building's public spaces, and the glass walls all revealed the importance of material understandings of the laboratory as a productive discursive space (Galison 1997; Galison and Thompson 1999; Knorr-Cetina 1992; Latour and Woolgar 1986; Rabinow 1997), with the laboratory's materiality as a central issue.

While Miku and Carlisle articulated how openness and transparency worked as part of a material, epistemological logic related to ideas about scientific modernity, globalization, and innovation, I also connect the concept of the open laboratory to the underlying notions of translation, mobility, and translatability. Translational science at the university worked through the creation of environments that compelled epistemic orientations on the parts of scientists toward openness and collaboration. I point toward the creation of a design in which an orientation toward openness reflected a design decision concerned with producing epistemic effects. According to these discourses, producing such knowledge environments essentially helps to shape the way inhabitants think and engage in scientific work. This project was not actually about mapping the effects themselves. While I discuss what I call *translational thinking* in a translational context, this project is not about thinking patterns. Instead, the focus is on how epistemologies and thinking patterns are *essential objects that are being designed for* in the creation of systems such as the CTSI website and in environments such as the "open" translational laboratory. In "The Design Process as a Critical Component of the Anthropology of Technology" (2001, 123), W. David Kingery revealed design's methodological solicitations:

> Design, the process of conceiving and visualizing an artifact, of forming a plan, of contriving an arrangement of parts in a device, a process or system is at the core of technological change. ... Design is required not only for the artifact but also for each of the production activity processes and use activity processes associated with a system of technology.

Thus, I focus on design and planning as a means to target envisioning and institutional planning in the creation of pathways such as those engineered into the backend of the CTSI website. The question of whether all subjects within these environments were actually changed, and the question of resistances (evidenced by Miku's curtain), rightly complicate any narratives that presume necessary outcomes for designs—especially when those outcomes involve the transformation of subjects. Nevertheless, in mapping a set of designs and logics, I hope to show how these designs are part of—and essentially work through—moral visions regarding TN and the translational project in general.

## 4.12  The Importance of Being Open

The CTSI website exhibits the sense of "open" as a hallmark in discourses about translational science. CTSI web-based tools enable sharing of data and resources across institutions and departments, and establishing connections with other stakeholders internal and external to UCSF. Additional web-based collaborative research tools such as Pfizer's, and others such as early brain-mapping technologies (Beaulieu 2002), facilitate the ability to more efficiently analyze data across sites, collaborate, and share knowledge. Simultaneously, the move toward open labs, cross-disciplinary research, and collaboration as concretized in the redesign of university laboratories and campus areas connects with what I have called *systems of epistemological contouring.*

While I had been thinking mainly in terms of the way such environments may be designed to compel translational thinking, Miku's lament caused me to think of the ways that openness operates toward different aims. Miku's visceral reaction to her open laboratory environment was marked by feelings of being exposed rather than "open." The narrative of "the open" neglected the reality of a multilayered positionality. The translational laboratory is open from some perspectives and closed in others. Discussions with Miku and Miu expose knowledge goals as they are manifested through designs. Thomas Gieryn's (2008, 798) study of Stanford University's award-winning biomedical engineering and science building revealed a similar design approach in university science buildings:

> Bright yellow "hotel benches" can be rolled anywhere if a scientist from outside gets teamed up temporarily with a Clark-anchored project. Nothing much is a secret at the Clark Center, since there are no doors to close off your lab space from other scientists, and even the hoi polloi can watch the action from those publicly accessible balconies.

In the new laboratory, openness and movability were central themes, along with the death of the historical world of scientific silos. In contrast to secrecy among scientists, openness would make science more accessible and coincides with a pedagogical philosophy that sees transparency as encouraging learning and innovation. For example, in May 2010, Texas-based Trinity University broke ground on a $100 million Center for Sciences and Innovation. In an interview with the *San Antonio Business Journal*, university president Dennis Ahlburg explained the use of glass in the building in terms of collaboration and demystification: "The exterior-facing glass walls will let people see inside classrooms as a way to demystify science and spark curiosity." In the other corner of the building is the Innovation Center known as The Cube, the center of what Ahlburg called the "entrepreneurial engine room" ("5Minutes—*San Antonio Business Journal*" 2011).

In contrast, "the closed" was narrated as part of a historical, anachronistic imaginary in science. I envision a closed, medieval science marked by formality, politicization, elitism, inequality, and exclusion (Galison 1999; Hannaway 1986). Open scientific structures and policies were therefore narrated against this particular depiction of a classical, prototypical science. Yet the rapacious growth of private patenting and licensing agreements, and the emergence of university lawsuits over ownership of scientific discoveries, have not only been well documented (Basken 2011; Mowery et al. 2004; Shapin 2008), but show a more complicated picture supporting a culture of both open and closed access. This dualism and rhetorical contradiction are similar to narratives of neoliberalism in which promises of openness via globalization are undermined by problems of global access (Biehl 2007; Collier 2005; Harvey 2005; Tsing 2004; Whyte et al. 2006).

Thus, ideas of "the open" and "the closed" intersect as a set of paradoxes. Tom Carlisle was eager to show me the brand-new open space conducive to sharing and collaboration. Yet increased demands regarding the development of intellectual property meant that he was reticent to share information about his laboratory's technical work or discoveries. Thus, the orientation of the knowledge worker has become both increasingly open and closed in translational science's environments.

From a market view, these partnership technologies were extensions of the market. To the extent that Carlisle's laboratory may find new means to make new partnerships using web-based partnering tools or consulting services such as those UCSF offers, openness may actually facilitate the meeting of ideas with capital. Transparency narratives function as self-fashioning vehicles for those involved in the translational system (ranging from investors to lab workers) and enable the laboratory to become a beacon for public-private partnerships accommodating an increasingly entrepreneurial university. An open lab also signals market openness.[13] The openness of translation, even in the symbolism of the concept of translation, invokes the evolution of the university.

In the introduction to their volume, Henry Etzkowitz and Andrew Webster offer a compelling articulation of the shift from the implicit knowledge transfer from universities to corporations toward a more explicit transfer model:

> Transfer of knowledge to industry was theoretically freely available through the literature. But in practice industry needed relationships with academic scientists to translate this knowledge into a usable form. This is one of the driving forces behind normative change in academic science. (Etzkowitz and Webster, in Etzkowitz et al. 1998, 3)

However, if the narratives of openness and sharing that animated translational science's knowledge environments (laboratories, webtools, new campus communities) were actually about an openness to the market, then TN must also include a means of *modulating* and managing such openness. I aim here to redirect attention. The question "Is modern science more open, collaborative, and transparent under translation?" may be the wrong question. Transparency and flexibility have always been understood to be inextricable from the power to enable openings and closures. Consider Georg Simmel's (1909, 67) thoughts about the functional polyphony and subsequent significance of the doorway:[14]

> Given the fact that the door creates a sort of hinge between the space of man and all that lies outside it, it overcomes the separation between the inside and the outside. Precisely because it can also be opened, its closing produces an even stronger sensation of separation from all that is outside this space than that produced by a mere undifferentiated wall. The wall is mute. But the door speaks.

Transparency became the articulation for market potential inasmuch as it became a means for it. The enunciation of openness works similarly in the case of the glass walls, where the discourse of openness belies realities of surveillance and control. As scholars have noted, discourses regarding transparency tend to produce epistemic effects (Collins 1999; Gieryn 1999, 2006, 2008; Kohler 2002; West and Sanders 2003) even when such designs do not necessarily lead to openness, collaboration, or innovation. Importantly, discourses about epistemic practice help to contour perceptions of openness, collaboration, and innovation in representing TN. How the design of spaces such as TN laboratories intersects with perception is reflected in Henri Lefebvre's (1991, 38) discussion of the semiotic representation of space:

> Conceptualized space, the space of scientists, planners, urbanists, technocrat subdividers and social engineers, as of a certain type of artist with a scientific bent—all of whom identify what is lived and what is perceived with what is conceived. … Conceptions of space tend, with certain exceptions … towards a system of verbal (and therefore intellectually worked out) signs.

Such discourses are especially important for translational science because of how TN *relies* on discourse to separate it from basic science. In other words, Miku's experiments, ongoing during the time of my visits, on neuronal death using animal models within the space of the TN laboratory were likely the same as the research on neuronal death using animal models several years earlier when it was not attached to something called "TN." Yet the consecration of the TN laboratory and its design are critically enmeshed in a new conceptual frame, the construction of TN as being *about health*. In many ways, we can think of the very design of the laboratory as part of this discourse about what TN is and how it functions in a broader knowledge economy.

## 4.13   Making the Moral Material: Environments for Translational Thinking and Perceiving

Thinking of design as a space of symbol making, objects such as architectural plans, engineering designs, and technologies reflect how TN exists in various registers. Software to support translational research shows the relationship between discourse, fantasy, and a kind of material actualization. In the case of TN, designing environments such as UCSF's CTSI webtools and Miku's redesigned TN laboratory constitute the material life of translational fantasies. Yet, to the extent that both the laboratory and webtools are also means by which university workers engage in scientific labor, one sees a fundamental manifestation of TN in the shaping of university knowledge workers and those who interact with them—such as venture capitalists and industry stakeholders.

In the way that designs for health interventions, economic regimes, and clinical trials implanted moral prescriptions (Biehl 2007; Jain 2006; Lakoff 2005; Petryna 2009; Whyte et al. 2006), they became a means to follow visions for intervening in the future. Similarly, designs reek of moral possibility, while also consecrating disorders befitting intervention. Similar to the characterizations of ethicists, designers change the "is" to an "ought." In this way, the designing of TN brings into view questions about how scientists, scholarly researchers, and the public ought to envision large-scale interventions in health while also providing an ethical archive of decisions about the (experimental, programmatic, conceptual) future of health.

Yet design does several additional things in the context of translational science. It targets the working out of *problems* thought to impede successful translation. An example of this may include the perceived problem of "academic silos" and the decision to create a building design that "solves" this problem by locating labs near each other. In this way, a set of designs become a symbolic space for the acting out of epistemic beliefs and the solving of "problems." Thus, this is how "ghost corridors" and plans for new science campuses become understood as actions done in the name of progress in health. In some ways, promises have merely rhetorical power, but designs enable the engineering of lasting environments.

However, though the university and the academic medical center are positioned as the de facto spaces for TSM, they represent only one element in the TSM system. As I learned during my fieldwork in Northern California, much of the "work" of translation actually happens behind closed doors in the sequestered spaces and networks of investors, entrepreneurs, and biopharmaceutical executives.

# 5   Bridging the "Valley of Death": Does Translational Research Nurture Innovation?

While a growing body of research has begun to explore the investor conference as an important site for scientific knowledge production (Eren Vural 2017; Lakoff 2005; Lock 2013; Sharp 2013), it is a critical site in the case of TN. Life science investor networking events are crucial spaces in which scientific tasks, projects, products, and agendas acquire meaning in a market context. Thinking about how things are made valuable (Birch 2017b; Doganova 2015), one can see how these events shape the sorts of industry discourses and market logics that render new partnerships, new technologies, or new research "valuable." Thus, importantly, the event itself functions as a critical site of "translation," and perhaps an apex in the "life cycle" of the translational pathway.

In the following section, I discuss how these conferences operate as de facto laboratories where scientific problems get worked out, relationships between investors and scientists were arranged, areas of investor interest are communicated to audiences of university scientists and entrepreneurs, and various kinds of neuroscience activity become understood as meaningful and valuable. A crucial site of "translation," these events constitute important theaters for the meeting of epistemology and value.

Yet as agendas for innovation in neuroscience and investments in biotechnology are outlined and articulated in these events, one sees broader implications: How is one to conceive of patients in these kinds of spaces? In this chapter, I propose that these sites were also theaters in which patient subjects are uniquely constituted via their biological parts—what I call "partial subjects." I also propose that TN relies on an "epistemology of parts," which tethers discussions about patient brains, cells, neurochemistries, and disease mechanisms to patient subjects, disease itself, and health. Additionally, these events were fascinating spaces where one witnesses how "problematic" scientific knowledge still ends up becoming commercialized and brought to the patient bedside in the form of technologies and diagnostics.

## 5.1   The Neurotechnology Investing and Partnering Conference

As already noted, the Neurotechnology Investing and Partnering Conference, is an annual (and sometimes semiannual) event in which university researchers, biotechnology startups, large pharmaceutical and device companies, and investors gather to make deals and set agendas regarding the future of brain-based medicine and technology. Yet, in its aim to connect "innovations" and innovators to resources, capital, and connections, the conferences also became a literal component of the machinery of translation itself. While there are several similar events nationally, the Neurotechnology Investing and Partnering Conference, sponsored by the Neurotechnology Industry Organization (NIO) and NeuroInsights, is the premier event for the sector.

The Neurotechnology Investing and Partnering Conference was also important because it helped make sense of the broader financial and market contexts within which TN operates and to which it connects. While science and technology studies (STS) scholars have taken laboratory studies to contexts *outside* of the laboratory—which anthropologist David Hess (2001) calls the "second generation" of STS—TN is oriented toward the interplay of the inside and the outside. Thus, for TN the "outside"—the issues of getting venture capital for biotech startups, navigating regulatory concerns, and connecting universities to for-profit partners—is very much a part of its intervention inasmuch as it is concerned with the petri dishes and modified animal models of the neuroscience laboratory.

Yet, in taking the mechanism of symbolic language (Black 1973) seriously, one could argue that the analogy of *the outside versus the inside* of the laboratory does not work at all in terms of translational science—at least not in the way that this distinction has been used throughout STS scholarship. In fact, much of the science of TN actually happened inside the Neurotechnology Investing and Partnering Conference, (see figure 5.1), a place where the brain and its parts accrued particular meaning, where scientific and financial practices were aligned, and where both experimental and regulatory problems alike were met with collective attempts at solutions.

**Figure 5.1**
Neurotechnology Investing and Partnering Conference is an annual event sponsored by NeuroInsights and the Neurotechnology Industry Organization—both managed by Zack and Casey Lynch.

It also became a context in which presuppositions and logics that would necessarily encumber and inform the work of brick-and-mortar science laboratories came to light. In many ways, the Neurotechnology Investing and Partnering Conference,seemed to function as TN's primary laboratory and site of key emergences.

During our first meeting, I asked Zack if he would allow me to attend the Neurotechnology Investing and Partnering Conference. He obliged and sent me a personal invitation. According to the email invitation, the attendees would include "CEOs, CSOs, CFOs, business development executives, non-profit leaders, corporate investors, venture capitalists, private equity investors, institutional investors, technology transfer experts, licensing executives."

When looking at this list, I did not see a reference to patients. I had somehow expected to see some representation of patients or patient advocacy organizations. An essential demographic that attends these events is the scientist/professor/entrepreneur looking to take his or her small biotechnology company to the next level and who is eager to create relationships with investors, licensing partners, pharmaceutical companies, and so on. The event's organizers had designed this invitation to entice this entrepreneurial contingent. Below the list of speakers appears the following crucial sentence: "Don't miss this excellent networking opportunity to discover partnering opportunities from across commercial neuroscience."

Two kinds of partnering happen at the Neurotechnology Investing and Partnering Conference. First, it facilitates the social partnering of entrepreneurs and their small biotechnology companies and potential investors. The conference provides opportunities for investors to meet companies on the verge of a significant discovery. In a 2013 interview in the *Life Sciences Report*, an influential commercial life science publication, Zack Lynch, director of NIO and cofounder of NeuroInsights, offered an explanation for why events such as these are important for entrepreneurs of early-stage neurotechnology companies, many of whom are university scientists:

> It is very important for entrepreneurs to know, in real time, who they might be able to partner with and how they might be able to get funding to bring products to market. Last year, VCs put more than $1.2B to work into the neurotech space, in more than 160 deals. Many of these companies have been speakers at our conferences over the years. We feel like we're playing a very important role in bringing the community together on an annual and global basis, to help galvanize innovation in the sector. (Quoted in Mack 2013)

There was a second kind of partnering that happens at the conference as well. These events became spaces for a material and semiotic partnering.[1] By this, I refer to the ways attendees reflect on and created-strategies (financial, market, regulatory) to deal with the messiness of biological material and to attach biological material to discourses and medical disorders.

These events are also ethnographically important because they are *closed*. By closed, I mean that they are "insider spaces" where certain kinds of product and knowledge anxieties are made explicit and performed out in the open. Among pharmaceutical and biotechnology companies, one finds triumphalist narratives that explain innovation and scientific decision-making. Even in some of my one-on-one discussions with pharmaceutical executives, I could detect a particular ideology—one that occurs, ironically, in the process of translating internal meanings to an outsider. Yet, in these conferences, conversations were less filtered. Here one could see explicit and often-contentious discussions about market demands and market insights, something often obscured within the "official" discourses of pharmaceutical companies, especially concerning health innovations.

Importantly, these conferences were crucial environments in which the articulation of scientific, informational, and market value *as it was being ironed out* by investors, scientists, pharmaceutical executives, and stakeholders took place. By the end of 2011, I came to understand that these spaces functioned not only as a key means of several kinds of problem solving, but also as critical spaces of scientific and material arrangement. It is in this latter sense that I began to think of this event as one of TN's primordial laboratories.

## 5.2   Investing in Risky Neurotechnology

The meaning of the term *neurotechnology* is as broad and unwieldy as that of the term *brain science* (Abi-Rached 2007). While neurotechnology is conceptually comparable to biotechnology, the latter refers to a large group of objects, disciplines, and fields loosely connected to biology (widely constituted), whereas neurotechnology is conceptually tied to the brain. So while neurotechnology retains specificity in terms of neuroscience, it encompasses a wide array of methods, diagnostics, and technologies. The term is also broad because the umbrella of the brain includes everything from blood-based diagnostic tools for Alzheimer's disease to digital games focused on helping US soldiers contend with posttraumatic stress. The ever-expanding set of phenomena to which brain science is tied means that many products and tools can be considered neurotechnologies. From weight-loss drinks to internet-based educational games, the neurotechnology umbrella (see figure 5.2) is intentionally large.

My first experience at the Neurotechnology Investing and Partnering Conference,was in the summer of 2010 in Boston. This particular event included a day focused on TN. The Neurotechnology Investing and Partnering Conference is always at high-end hotels, in San Francisco or Boston. There are occasional fringe events in international locations or adjacent

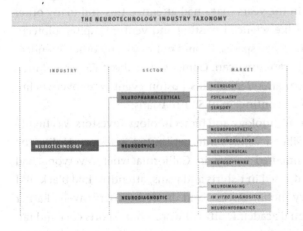

**Figure 5.2**
The neurotechnology industry taxonomy. Courtesy of NeuroInsights.

professional-society meetings such as the Society for Neuroscience's annual meeting.

The May 2010 meeting was held at the Westin Boston Waterfront. The standard registration cost was around $1,200 for most attendees, with slight discounts for nonprofit attendees. The Westin Boston is a beautiful hotel. In the lobby a tall cascading wall of dark wood makes the space feel dark. The hotel's gently curved ceiling meets a tall, slanted expanse of windows that look out to Boston Harbor. Near the entrance, a procession of hotel attendants greeted fleets of black town cars. Attendees quickly exited the cars and immediately went to the registration desk. The event included a mix of people, but a large contingent were investors, biotechnology-company executives, and pharmaceutical executives.

The main meeting events all occurred in the main banquet hall on the mezzanine floor. Inside the banquet hall, there were rows of tables and chairs all facing the raised stage area. This area included a central podium, flanked on both sides by tables for panelists. As I sat down, I noticed a very small shiny silver pin. Placed at every table in front of every seat was a metal pin molded in the shape of a brain. The pin was attached to a card that contained a single phrase, "*Give the brain a voice.*" Only later would I realize how prescient and appropriate this charge would be.

Having arrived early, I chose to sit in the back both to gain a perspective on all of the activity and perhaps as a way to recede into the background. I certainly stood out: in the years that I had attended this and many other events in the world of life science investing and venture capital, biotechnology, and neurotechnology spaces, I could not recall any other attendees who appeared to be African American. Composed of about 75 percent men and 25 percent women, attendees at this Boston event were overwhelmingly white, with a small contingent of South Asians.

The culture of neurotechnology and biotechnology investors was highly distinct from that of Silicon Valley's internet startup investors. Unlike the youthful technology-oriented events in California where everyone had Apple computers and dressed in t-shirts and jeans, attendees had black IBM laptops and BlackBerry mobile devices and dressed conservatively. Rather than choosing traditional academic attire, I wore a blue sports coat and tan slacks, likely a result of my interpretation (and probable conflation) of a wardrobe befitting Wall Street and/or members of a Princeton eating club.

Despite my clothing, I was still an anomaly. Frequent, surreptitious glances were not uncommon. However, the subsequent scan of my Princeton name tag seemed to neutralize any externally discernible lingering interest. Among a sea of university researchers and entrepreneurs, perhaps it offered a quick narrative that explained my presence. This event was one in which name tags mattered. Affiliations were in a larger font than one's name. Presenter biographies were lengthy and sparkling: there was little variation among investors' very similar elite pedigrees. About 90 percent of all of the investor panelists and speakers at the event, whose written bios presented their education credentials, listed degrees from an elite and/or Ivy League school. Several of the attendees who had degrees from nonelite schools mentioned their degree and did not mention the institution. Presenting scientists, most of whom were also entrepreneurs and were attending to share information about their startups, displayed a greater educational variety, including a smattering of scientists from nearby Harvard and MIT.

On the first day of the Boston conference, I happened to sit next to the investor that I mentioned earlier, Michael Hall, with whom I met in his San Francisco office overlooking the Presidio. In this, our first meeting, he looked at me and glanced at my name tag—and I his. His well-regarded venture capital firm was one of a small circle of firms that regularly had a presence at these events. I introduced myself. I explained to him what I was doing—that I was an anthropologist and that I was working on learning more about neurotechnologies and TN in general.

"Hmm, that's interesting. ... Sounds cool," he said. I asked him about his interest in this event. Michael was a relatively young partner at a venture capital firm that specializes in biotechnologies located in the southern portion of the San Francisco peninsula, not far from Palo Alto and Stanford University. "But they let me keep an office in San Francisco," he said. His firm, a darling in Silicon Valley's tight biotechnology community, was involved early on in several medications that, in his words, "had amazing exits." In the world of venture capital, the primary concern for every firm that invests in a startup company is the issue of a successful *exit*—which means that the company receiving investment is eventually acquired for a significant sum. Potential acquirers include pharmaceutical companies, a private equity company, or a large holding company. In the best scenario, the startup company will grow and eventually go public via an initial public offering (IPO).

In either an acquisition or an IPO, a firm's initial investment can yield a substantial profit. For example, Michael's firm was an early investor in a Boston-based oncology startup. After several years a large pharmaceutical company purchased it for over $500 million. As a result, that firm's investment fund yielded a large profit that quarter, which in turn attracted additional investment. According to Michael, the firm's principals made seven-figure incomes that year from management fees. The profit from that one exit was enough to make up for the failures of other startups that were part of the firm's investment portfolio. "We are always looking for that *one* company," he said, holding up one finger. A glint in his eye, I could sense his passion and anxiety.

I asked Michael additional questions about his experience at these events. "I've attended this event for several years now," he said. He then expounded on the role of social connection in these events and the larger world of venture capital and investment circles:

> I come to these events for a couple of reasons. One thing is that it's important for me to be well connected and to nurture my connections because a lot of times when I'm investing with people, I may co-invest with other investors and other VCs [venture capital firms]. So, the relationships that I've built, I need to maintain. And I've built a lot of 'em. But I also go to these events to find out what's going on in the field, what new stuff is coming out, what people are thinking in terms of Neurotech and Pharma. I also like to see how scientists are framing certain problems. I also like to see what other investors are doing. Sometimes it's difficult to get to have one-on-one conversations with other investors. These conferences give me a glimpse of where other investors are. And of course, you know I also am looking for deals. (Interview 14, May 20, 2010)

He explained that he attended these events "just to hang out" and perhaps most importantly, to see if he "can get deal flow." I asked him to talk more about "deal flow." While it was clear that these events were important for small biotech companies as they sought investors, capital, advice, or key relationships, I could see that these relationships are part of the *product* of the event. Yet I was also interested in investor perspectives to understand the underlying market logics that animate investments in neurotechnology companies. Michael explained what he was looking for concerning possible investments:

So, deal flow is basically a phrase that we VCs use to talk about the possible companies with good ideas. They [biotech startups] are all looking for money. But finding novel ideas, new models, new things to invest in is not easy. I see probably about 40 different ideas a quarter and I may fund two or three of them. The problem, though, is that we are always trying to get a piece of the hot new thing and to not miss it. And I know a lot of startups think that we don't really care about the startup exhibits [presentations to investors], but we really do. I'm always looking for the next big thing. Right now devices are really hot. So, I'm always looking for device verticals. I'm also always checking up on the Big Pharma panels. I used to be part of that world and have lots of friends who still are. I'm always looking to see where Pharma is going and if that can help our current portfolio of investments. We always need for our portfolio companies to have good and successful exits. So, that also means understanding where the science is right now, what the consensus [is] about certain disorders. Here I learn both about the industry and the science all in one. This is a major part of my education around CNS each year in terms of progress, evolutions, challenges. (Interview 14, May 20, 2010)

The first panel of the morning in the May 2010 conference focused on the current state of investment/investing in neurotechnology. Panelists consisted of five investors from venture capital firms. Of all of the events, this panel was the best attended. Each panelist, a key decision maker in his respective firm, outlined what firms were generally looking for when investing. Some hinted at the interests/niches of their specific firms: "We're looking for good opportunities, good *teams*," said one investor. Another investor offered a useful indication of his firm's shift toward investing in less risky propositions:

The majority of investments in new ventures nowadays happen at or after phase two. We're looking for good science and if there is inefficiency in market valuation after market data. We try to mitigate our risk and prefer to invest after some of the clinical risk has passed, until there is better efficiency in the market and the company is better valued.

For this investor, involved in a later-stage firm, his firm moved away from considering companies in the earliest stage of research and development. At the end of the panel, there was a question-and-answer session. An audience member asked what qualities each investor's firm looked for specifically when evaluating startups: "Venture capitalists get a lot of inquiries," Justin Long said. Long then listed things that he and his firm are not interested in: "I wouldn't look at epilepsy," he said, drawing his finger across his throat to symbolize the problems associated with epilepsy research. One investor said smugly, "The best proxy for a venture capitalist is to look at what companies they have." Another investor added, "Uh, I would look at neurodegeneration, where there is high unmet need and a known mechanism [of action]. I'd look at schizophrenia."

Although each investor articulated widely different product and vertical interests and aversions, a tone of caution permeated the panel and the conversation. While the panel's title, "Investing in Neurotech," seemed innocuous, the conversation quickly displayed a deep crisis in the industry. This crisis was about the lack of new drugs coming in the pipeline, drug patent expirations, and a failure to create safe and effective psychopharmaceuticals. Few were explicit about the crisis during conversations outside of this event. However, inside the event, it informed the panels, the conversations, the audience questions, and even the written conference material. In the description of an afternoon panel on new business models in neurotechnology, one sees clearly articulated the way that risk and time tolerance are implicated in investors' shifts:

> With venture capital steering away from long term and risky investments, neurosoftware, neurofeedback, neuromarketing and other new business models represent far lower risk and faster times to market. What winning strategies are emerging in brain fitness, disease management, neuroplasticity, advertising and even sports performance?

Several of the panelists spoke about the problems in translating brain science and the subsequent risks. According to one panelist, "Translational risks make it difficult for big biotech companies to invest, but people are trying to use TN to forge new understandings." One panelist insisted that even though the latest neuroscience research is only providing *information* at this point, this information is still valuable: "This is still important," he implored the audience. In carefully crafted language, panelists acknowledged the shift away from neuroscience by Big Pharma because of a multitude of risks, including clinical-trial risk, problems with pharmacological efficacy, and a lack of neurobiological understanding. As scholars have noted (Hopkins et al. 2013; Styhre 2015), such a retreat by investors or Big Pharma necessarily impacts investor decision making. "Everyone is scared now," Michael later said to me. Always witty and curt, he added, "Things change quickly."

Yet some panelists exhorted others to try to maintain a view of neuroscience's importance, especially concerning the dire need for increased neuroscientific understanding. The pleas for sustained faith by some panelists revealed hues of dignified desperation. Eventually, the panel moderator brought up the issue of risk tolerance among neurotechnology's investors: "There *is* big fear that the big companies are getting out of neuroscience, especially in psych [psychiatric drugs]. This sentiment reflects the evolution away from psychiatric drugs, once a boon of the pharmaceutical industry" (his emphasis).

One of the panelists quickly responded: "Big Pharma is evolving toward the things that they do better—which is big [phase] three trials, and the things where you need additional soldiers and this leaves more opportunities for us." The "us" to whom the panelist referred is the cadre of venture capital firms as well as individual investors (some of these firm executives also invest their own wealth in deals as private individuals). It also means that these investors believe that new opportunities emerge in the retreat of large pharmaceutical companies from psychopharmaceuticals and away from the discovery process in general. Lastly, it provided a glimpse into the evolution of the pharmaceutical industry's strategies—toward those in which their "soldiers" can prove advantageous. That is, it signaled the dissolution of the traditional in-house R&D pipeline and a movement toward the outsourcing of various parts of that process.

Another panelist, Dr. Simon, a Yale MD/MBA and his firm's managing partner, made the risk issue more explicit. According to Simon, "the risk is *huge* in neuro" (his emphasis). He continued, gesturing emphatically, "You don't have the models." Several high-impact research articles began to emerge starting in 2006 discussing the hidden, unspoken problem of pharmaceutical efficacy. One meta-analysis published in 2008 made a bold claim that antidepressants generally aren't effective above and beyond placebo among non-severe persons (Kirsch, et al 2008)—a claim that met great criticism and controversy. Nevertheless, such an admission by an investor in neurotechnologies was surprising, especially in a context where optimistic narratives about pipelines, progress, and "innovation" are critical elements in maintaining investment (Andersson et al. 2010).

Simon went on to say that while big companies are certainly interested in these products, they don't want to be involved at the beginning stage. He recounted some of the many problems specific to neuroscience—including the growth of quick generics, the high failure rate in phase three trials, and the daunting costs of neuroscience research: "There is significant need in

**Figure 5.3**
A Europe-focused Neurotechnology Investing and Partnering Conference in Helsinki

psych, in neurology as well, but Pharma has a big interest in neurotechnology." According to Simon, "Psych is very last year."

His statement about psych prompted a thought about health. Up until this point, I hadn't been explicitly thinking about health, but it was jarring to think about mental disorders in terms of trendiness, especially in the context of semipublic discussions about health. Of course, arguably, this conversation and the larger conference were not principally *about* health. In fact, one could argue that these conferences, which take place annually in major centers all over the world (figure 5.3), are about something else entirely. Regarding a recent risk his firm took, Simon recounted in a moment of humor, "We've bravely … or foolishly … created a new company in psychiatry." The audience laughed at his turn of phrase.

The investors' clear admission about the lack of models, pharmaceutical companies' abandonment of psychiatric product areas, and the move toward new investment trends showed the impact of market logics on companies' R&D innovation strategies. This conversation also illuminated several possible neuropsychiatric futures—for example, the implications of this massive deinvestment for the practice of psychiatry; how this shift might impact those living in the sticky webs of psychiatric diagnosis; and for those already diagnosed, what to make of Simon's refrain, "You don't have the models." What about the lives being lived inside those models right now?

The panel alluded to the importance and opportunities that emerge around risk for postcollapse, postboon investors. In thinking about the risks attached to the "early stages" of neuroscience research—including the stage where, ostensibly, messy foundational and underlying questions and models are supposed to be ironed out—we see a key intertwining of epistemic and financial risk. We also see the terrain on which the outside—here the admissions about the problems with "models" in neuropsychiatry and the risk attached to the general sector—begins to produce contexts that condition the inside, including the questions that may get assigned to TN laboratories, and a clear set of meanings that are being attached to recent neuroscience research. For Simon to proclaim that "you don't have the models" is an interesting instance of a public "reading" of scientific work by a powerful stakeholder. Here the "insides" and "outsides" of TN come into clear view.

## 5.3  Mechanisms of Action

During the same panel, an impromptu conversation between audience members and panelists turned to the problem of scientific evidence—an issue that was especially salient for investors and VCs. The conversation focused on a discussion about biology in the form of the mechanism of action (MoA). For Simon, MoA was an essential component of his firm's decision making regarding investments in new startups: "We don't invest if there isn't knowledge about what the mechanism of action is. And there are many tempting things that we find in which we want to invest. So, novelty also has to be part of something that has to be demonstrated."

Wherever industry insiders gathered, few terms emerged as frequently as MoA. Sometimes referred to as mechanism of drug action, MoA is essentially the particular biochemical interaction created through a pharmacological drug substance (Spratto and Woods 2013). It constitutes a sort of biochemical roadmap that explains its effect on the body. It requires an understanding of a set of molecular targets on which a pharmacological drug substance works.

Thus, given its pharmacological focus, the translational process is usually organized around identification of a specific molecular target (Garcia-Rill 2012. Out of a molecular target, an entire research agenda is designed with the hope that it can be translated into a biomedical solution. The importance of the MoA for biomedicine reveals the underlying *pharmacoepistemology* at work in the way that we understand not only the structuring of research agendas, but also the therapeutic logic of pharmaceuticals.

Importantly, this epistemology consequently animates the social—and specifically epistemological—construction of disease. As medical anthropologists have observed, the notion of MoA has helped to engineer the foundation of Western biomedical research since the 1950s: "Strictly controlled research to demonstrate pharmaceutical efficacy and mechanism had greatly influenced the way biomedicine conducts research today—that is, by privileging the placebo-controlled RCT [randomized control trial] as the 'gold standard' for demonstrating efficacy" (Thompson, Ritenbaugh, and Nichter 2009, 115). MoA constitutes a biochemical bridge that connects human biology to pharmacology and pharmacology to disease claims.

Apart from its definition, it is compelling even in its phraseology, which confers an aura of surety and authority. Yet, for investors and for laboratory scientists, MoA—the term and the question about its importance—indexes the anxieties and possibilities that surround the research process in the search for novel scientific facts that can be turned into medicines and products. MoA is a key unit of scientific and market value.

During this panel, I began to see how MoA constitutes a sphere of corporate, social, biological, pharmacological and investor trust. MoA suggests that there is a clear factual and biological basis not only for understanding drug interactions, but also for understanding the underlying mechanisms involved in the disease state itself, or its symptoms. Given its importance in randomized control trials and for notions of pharmaceutical efficacy, MoA constitutes a kind of epistemic gold. In other words, in a context of great scientific uncertainty, MoA ushers in the sense that what one has found is scientifically real.

Yet it is not only a proxy for investor confidence, here consecrated in Simon's confident claims about its necessity in the world of investments. It is also, perhaps paradoxically, a proxy for scientific anxiety. This anxiety has loomed large across biopharmaceutical sectors. A controversial (and contested) 2017 article in The Guardian proclaimed, "Of 48 cancer drugs approved between 2009–2013, 57% of uses showed no benefits and some benefits were 'clinically meaningless'" (Davis 2017). Similar proclamations about "benefit" were made in the press about antidepressants and other medications. The debate over MoA indexed genuine anxieties over the terms of biological and pharmacological truth, even though finding the MoA was still no guarantee of a safe and effective therapy. To have a conversation about MoA was to have a conversation about the problematic nature of much of the knowledge that underlies pharmaceutical innovation and even the heuristics (such as MoA) used to "deproblematize" that knowledge. Lastly, it is also imperative to acknowledge that many neuropsychiatric solutions found in the twentieth century were epiphenomena—unexpected effects discovered from drugs developed for other purposes. Myriad scholarly histories detail these facts and thus are necessary for understanding part of how MoA has become a touch point for neuropsychiatric anxiety and history.

Visibly interested in the MoA question, the moderator asked if MoA is also important for investing in devices. By devices, the moderator was talking about technologies like the deep brain stimulation (DBS) device, a special kind of brain technology in which a pacemaker-like device delivers electrical stimulation through electrodes implanted in the brain itself. In response to the device-related diagnostics question, one investor, an MD, responded: "In many cases, we haven't understood the mechanism of action in these [clinical] trials as we should have." He continued, "In devices, the bigger challenge is the reimbursement area. I think it's important in devices since there is a large upfront cost compared to biopharma; you really have to show an improvement in disease outcome, but also a reduction in system cost to see rapid reimbursement potential." Here the risk for devices is treated differently than that accorded to biopharmaceuticals.

A woman in the audience stood up and commented:

> We should be a little less obsessed with mechanisms of action. In terms of devices, we don't have a clue about how deep brain stimulation works ... not a clue ... but it works now ... [her emphasis]. Yes, mechanism is nice. But let's not throw out the idea where we don't know the mechanism. Biologically, we know a little bit about the receptor in the drugs and [yet] there are lots of impacts in the brain.

Simon responded that his firm "had some successes" with an approach focused on investing in companies where there is a proven mechanism of action, "and so that's why there is such an important check box for us," he ended. The audience member's admission—which came across as more of a collective acknowledgment than merely a statement of personal knowledge—elucidated a VC perspective on biological ambiguities embedded in devices such as DBS. Concerns regarding patient perspectives and realities did not color this exchange, but this discussion was also about an ambiguity over biological material.

This was not only a unique problem for investors and the researchers in the audience; MoA is also a regulatory problem—a problem made evident in the battles that companies wage with the FDA. Given the power of the placebo effect, MoA is also a way to demarcate actual pharmacological effects from the effects of a placebo response. One panelist went straight to that point by saying, "In neuro, the placebo effect is quite large. We now know that X drug works and so now have to figure out the regulatory and the market strategy." This metaconversation about the importance of the MoA is much more about a larger *working ambiguity* involved in biological sciences.

In a different panel, the back-and-forth over MoA delved more clearly into the question of the molecular basis for biomedical intervention. At one point, the panel discussant, a director at a life science advisory firm, commented on the new emphasis on this kind of molecular biology–focused scientific knowledge. According to him, "This new approach is interesting especially in placing a premium on molecular biology knowledge." This signaled a shift—even if only within a single advisory firm—toward pushing for more molecular biology knowledge or approaches. As an advisory firm that provides strategic consulting to companies and investors alike about how to profit/grow, this ethos could translate into consultations with biotechnology companies. This was the emergence of distinct scientific direction and at the very least, scientific possibilities, in much the way that one would observe such emergences and possibilities in a traditional, brick-and-mortar laboratory.

Then Chris Marshall, a Princeton PhD and chief science advisor at a Canadian pharmaceutical company, offered a caution about having too much knowledge. According to Marshall, "Knowing everything that happens—having the mechanism of action—actually makes things complicated. The FDA doesn't have a requirement that you present the mechanism of action." Another panelist offered a useful distinction, one that set off an exchange about how neuroscience had changed and about how it is unlike investigations of other disease categories: "I don't think it's a big deal. I think it [the MoA] is nice to have. It's certainly not necessary. If there is good evidence that it's safe and effective, that's it. The productivity of neuroscience has declined in terms of specificity. It may work for cancer but not for neuroscience." This admission about specificity—and thus about the diminished depth or "quality" of contemporary neuroscience—was telling.

Mary Holborn, vice president at a prominent US pharmaceutical company, responded, "We're agnostic to mechanism." She continued, "Some of our most interesting things are where we don't know it." Holborn's sentiments matter because she directs her company's acquisition of companies and licensing of intellectual property. Given that this pharmaceutical company is involved with bringing emerging technologies to market and her decisions about what to focus on had likely influenced companies and university scientists in the audience working on ideas, this scientific position is meaningful in order to answer the question about how technologies and their material ambiguities—models, assumptions, foreclosures—are brought to the patient bedside. After more back-and-forth, Chris Marshall, the chief science advisor of the Canadian pharmaceutical company, offered an important admission that crescendoed throughout the ballroom. Uttered in a tone somewhere amid incredulity, gratitude, and joy, Marshall disagreed with the necessity of deeper knowledge. As he put it, "Not knowing our molecular target is what got us to where we are *today*! That was a pretty productive way to do drug discovery!" (original emphasis).

For these two representatives of major pharmaceutical companies, scientific ambiguity, especially around molecular targets and their mechanisms, was not necessarily a problem in bringing products to market. This indifference to ambiguity marked a clear distinction from the stance articulated by the investors, who inevitably absorb a significant amount of risk in early-stage biotechnology investment (Brown 2003), and very much want mechanism knowledge. Observing this bifurcation revealed the various positionalities regarding risk, though it was curious that pharmaceutical companies were more amenable to dispensing with this knowledge. In the life cycle of a neuroscience discovery, where and when do its constituent facts matter and for whom? One astute audience member asked a question about the public. She asked how new innovations are being thought of in terms of their potential to help patients. Marshall's response seemed accidentally unambiguous, if slightly dysphonic and robotic: "Health economics plays a big role in our [business] development teams at [her pharmaceutical company]. And the FDA and payer environment wants to see clinically meaningful benefit."

**Figure 5.4**
The 2011 Neurotechnology Investing and Partnering Conference at the St. Regis Hotel in San Francisco. A group of panelists are sitting on the makeshift stage.

The setup of these events (see figure 5.4) meant that the discussions among panelists and this back-and-forth between panelists and audience members took on a dramaturgical dimension. Perched at the front of the room and flanked by large screens on both sides, the panelists were more or less literally on the spot. Of course, Marshall's response was strange. Wouldn't the pharmaceutical company itself also be interested in "clinically meaningful benefit"? Perhaps it goes without saying, but juxtaposed with the first sentence about health economics, Marshall's statement dismissed clinically meaningful benefit as a necessary hurdle among many for business development teams.

One audience member quickly approached the microphone. The woman's speech was slow and weighty. It was of a noticeably different tempo than the swift rhythms that marked the event up until this point. She said, "Remember, we're all patients *too*" (original emphasis). She then sat down. With a sheepish smile, the panelists nodded as if to suggest agreement, but the exchange ended there.

This comment was compelling. First, Marshall never actually mentioned patients. His discussion did not explicitly focus on patients at all. It was also interesting because, in its concision, his statement presumed that everyone in the audience and on the panel would necessarily understand its relevance, as if they were all thinking what he was thinking. The woman's response from the audience may have been a way of saying that clinically meaningful benefit actually matters; that it is not merely a perfunctory check box during business development processes. It was perhaps a way of saying that we, too—as both professionals in this industry and as human beings—have a vested interest in pharmaceuticals that *actually work*.

The claim that "we're all patients *too*" was a jarring reminder of the back-grounding of patients during this event. After some time, I had begun to think that patients were simply missing within these discourses. But that was wrong: patients were present, though not in the ways that I had antic-ipated. Some presentations included generic pictures of faces that were ostensibly those of patients. Occasional discursive gestures to health and patients were fleeting and cursory. One audience member, who worked for an advocacy organization, brought up the issue of patients, but the panelist responses to her questions incited responses about management and FDA regulation. Patients were also implicated in the pieces and slivers of images of cells, fMRI brain images, and data scans that filled the presentations and event brochures. The patient subject was a pixelated one—a subject that was both missing and yet inextricably implicated. People were indeed pre-sent, but they were there in part or in *parts*. Patients were dissected in this space and discussions about patients were about them in their parts. This was, indeed, as much a laboratory as it was a conference.

## 5.4   The Socialities of Translation

Nevertheless, these events went beyond mere theaters of discourse and sci-entific negotiation. These spaces were also a crucial avenue through which new relationships were born, forged, renewed, and emboldened. Built into the very design of these events lay promises of opportunities, connections, and partnering. Consequently, discussions in the main hall were not the only conversations that mattered. Many significant conversations and con-nections occurred in spaces hidden from view.

The 2011 Neurotechnology Investing and Partnering Conference was different from its predecessor in Boston. Held at San Francisco's St. Regis Hotel, this event space felt much more intimate and exclusive. The St. Regis is a noticeably more designed hotel than the Westin Boston Waterfront. Often, the conferences in San Francisco are been held at this venue. The Boston event has since switched to the Boston Ritz Carlton.

The San Francisco St. Regis (see figure 5.5) is both a hotel and a luxury condominium building. A towering, modern, glassy structure, the ground floor is flanked on one side by the relatively new Museum of the African Diaspora, known as MoAD—a modern and sparse space more reminiscent of a gallery than an actual museum. On the other side of the building,

**Figure 5.5**
The lobby of the St. Regis Hotel in San Francisco looks out onto the hotel's private driveway, which connects to the adjacent city streets. Surrounded by glass, the lobby provides a view out to downtown.

a porte cochere gently indents the side of building to accommodate the hotel's private driveway. In the middle of one of the most frenetic and congested parts of downtown San Francisco, the hotel feels recessed from both the street and the city at large. Surrounded by glass on two sides, the ground-floor lobby offers a clear view to the cavalcade of town cars that drop off passengers. On the ground floor, there was a prominently placed bar, a loungelike space with light-colored, modern, lush furniture. The elements of the space stand out. The lounge space is chic, severe, and luxurious. Adjacent to the bar area, the lounge also includes nine or ten seating arrangements. As I arrived, I recognized several faces from the Boston event. Already with cocktails in hand, a group of men I recognized as panelists at the 2010 event were sitting in the corner of the lounge area on the ground floor. The space was so suitable for socializing that I began to take note of the ambience of the lounge and its design, to watch how sumptuous club chairs and dim lighting seemed to coalesce with sudden intimacies and social lubrication. The registration table was a few floors up. I grabbed my

**Figure 5.6**
The St. Regis mezzanine. The hallway serves dual purposes. It is both an entry to the main event space on the right and also functions as an immediate, external social space away from the main events. During social hours, the hotel staff brings out cocktail tables, and the space became dedicated to mingling.

name tag and immediately explored the space. On the same floor as the registration section lay a large, brightly lit hallway area (see figure 5.6). Surrounded by glass on two sides, the area served as the entrance to the main hall where the sessions occurred.

One significant function of this event was to facilitate the social relationships and networks at the heart of TN. The movement of a finding from a scientist's laboratory bench to its actualization within a biotech firm requires a host of social formations and connections. Michael Hall had previously indicated the varied ways that social interactions work in the "translation" from the laboratory to a market-ready product: from nurturing connections already made, to keeping abreast of what's going on in neuro and investing, to looking for deals. But here in San Francisco, I began to wonder how sociality functions for everyone else: How do new entrepreneurs navigate these social spaces? What about pharmaceutical-company executives? Beyond already existing relationships, how do individuals form new relationships? What do social relations mean and do in the worlds of TN? I wanted to know more about the conversations that happened right outside of the main event space and in adjacent meeting areas.

On the first day of the San Francisco event, the importance of social connection became suddenly clear during the presentation of an unusually loud and colorful presenter, Leonard Elliot, a rotund man with big expressive blue eyes. Quick to offer a joke and a grin, his smile is distinguished by unusual intensity. As a vice president for business development and research at a major medical device company, he is charged with helping the company create alliances with partners—smaller firms, biotechnology companies, universities, or even large pharmaceutical companies—where alliances may help his company acquire, access, or develop new products and markets. These alliances can come in the form of licensing agreements and can even take the form of an acquisition of a firm or its intellectual property. For many venture capital investors and startup entrepreneurs, an acquisition is a good "exit" and thus makes Leonard a key person to know.

In fact, as he outlined his company's partnering strategy in his presentation, he began with a remark about the importance of sociality, which he called altruism. He began his presentation with a long pause. Surveying the audience, he started his speech with a drama that is uncommon at these events:

> Be altruistic ... "What?!" you're thinking, "You're not supposed to be altruistic!" But this meeting shows how important altruism is! More than half of you [he gestures toward the audience and smiles] are my friends, and more will hopefully become [so]. Be altruistic, or you will not find the relationships that are necessary to create alliances.

Leonard's directive to be "altruistic" was interesting. Moreover, the audience members in the room numbered 400 or so at any one time, as a stream of people entered and left the main event space. Was he serious that half or more of the audience were his friends?

During the break between sessions, I went out into the hallway right outside the venue. In a bit of luck, I spotted Michael Hall thumbing around on his smartphone in the seating area adjacent to the main event space. It had been a while since we had seen each other. As we exchanged small talk, Michael would catch various other attendees' eyes and other participants would gently touch his shoulder as they passed. A flurry of microconversations permeated our conversation. I asked Michael about Leonard's statement about being friends with half or more of the audience. Michael had not seen Leonard's presentation but agreed that Leonard was likely to know much of the audience.

"He's *super*connected," Michael said. "He's definitely someone you want to know. I tried to get him to be an advisor for one of my portfolio companies." A ball of energy, Michael stroked the side of his hair every so often. He continued, "He's so busy, but he is just the most connected person in devices. But you know, [Leonard's company] has been on a rampage lately. It's probably true that he knows half the people in that room or more." Michael explained that Leonard has been at the helm of several high-profile acquisitions, so small startup companies were always trying to talk to him. According to Michael, companies do best if they first get an introduction through Zack.

I asked Michael about Leonard's comments about altruism. Seated in his chair, Michael shifted his body and straightened his clothes. Softly clearing his throat, he turned to me. "Well, one thing you need to know is that everybody's hungry," he said, as his eyes eventually met mine. He continued:

> For startups, for one thing, they are at the bottom of the totem pole in one sense because they don't have anything. I mean they only have a few compounds or some other thing that they think that they've found. But they are at a disadvantage and yet Pharma is in big trouble right now. I mean, it might not look like it, but investors are up in arms, Pharma is closing their neuroscience arms left and right, and so far they have been just skirting along, you know. Leonard works in devices, which is the hot thing right now. Delivering drugs through devices is a unique angle, especially because you can wrap services around it. I say all this to say that while startups are the little guys, the big guys still need them. Pharma especially. The medical device companies have been eating up small biotech firms like crazy. That's why my firm really tries to find device people. But small device companies are hard to start. It takes lots of money. So a lot of times a professor might come up to me just with an idea. You know ideas aren't as easy to claim as devices. So then the risk is that if you tell the wrong person your idea, [any company] may very well just take that shit. I mean just take it. So a lot of times Leonard wants to have conversations, but some of the little guys are scared to talk to him. In these spaces, info is key. But, at the same time, every startup founder thinks he's some Einstein or something. (Interview 20, May 21, 2010)

"Everybody thinks they're Genentech," he said, referencing the billion-dollar biotechnology company that began out of a scientist's laboratory at UCSF (Hughes 2011).

Michael's impromptu proverb, "Everybody's hungry," adeptly sums up the tensions and risks at work in sharing information, a process both necessary and risky. Moreover, how much to share and with whom is not obvious. For entrepreneurs, for example, the task of finding the right people with whom to share is not easy. A person's placement on the "totem pole" determines their vulnerability.

It is likely that Leonard's use of the term *altruism* was a nuanced way of saying that to remain competitive, to learn about new business opportunities, and to assess new ideas and movements in the field of medical devices, social connections were vital. Yet, as crucial as social connections were, it was information that was the currency of exchange. Leonard, Michael, investors, Pharma executives, Zack, and the entrepreneurs—they all needed people *to share* with them.

While the sociality at work in this event was clear, these spaces harbored vital *financial* communities. It turned out that a large number of the attendees knew each other and also had strategic financial ties and relationships to each other. Startup entrepreneurs, for example, needed financial investors, and life science investors needed to find that one business that will create enormous financial returns. Zack leveraged his network to be able to find clients for his advisory services and market research products, and pharmaceutical companies needed these relationships to find potential acquisitions and/or to learn new strategies that will lead them to increased profits or minimized losses. The concentric circles of financial dependence and relevance shed a different light on the webs of social connection at work. Interested in the specific social connections among the group, I asked Michael to do a quick search on his LinkedIn profile. He confirmed that he and Leonard, with whom he was connected, shared a larger concentric circle of contacts—people who were also at this event. LinkedIn revealed that this world was a small one built on a vibrant network of connections.

Nevertheless, the risk that a company or investor could effectively steal a small startup's idea presents a challenge for entrepreneurs. I imagined that secrecy could, at times, also benefit investors like Michael, who could be incentivized to take a novel idea and give it to one of his firm's portfolio companies. I asked him about that possibility. With perfect confidence and lightning speed, he responded, smiling, "Naw, I'm a good guy." Then he winked. Tepidly, I smiled back, not knowing if he was joking.

Michael said that pharmaceutical companies have partnership interests too and that these interests are similar to those of the large device companies, such as Leonard's, for which internal innovations had come to a standstill. During a break between sessions, he introduced me to one of his contacts, a Harvard scientist named Simon Chiang. The two had met at the Boston event in 2010. Michael and Simon quickly caught up before the conversation delved into the crisis at biopharmaceutical companies, especially in the area of CNS disorders, and the subsequent need for these companies to create partnerships:

> **Simon**: Yeah, I had a couple of meetings with [a large US pharmaceutical company], and they were really interested in what I was doing. Their VP was telling me how the only way that they've been able to get new ideas has been by partnering, and that's a real focus this year. I see them all the time around our part of campus. I feel like they should just open up an office inside Harvard [he chuckles].
> **Michael**: Yep! That would be funny. You know, we've been doing well because they [pharmaceutical companies] are so thirsty for ideas. It's crazy because they are still raking in billions every year.
> **Simon**: I know. I'm always thinking, how is Pharma so poor, and yet still pulling in billions? You know my sister is in a science office at a biotech firm in Cambridge, right? And she says that part of the whole deal [the anxiety around revenues] is from investors.
> **Michael**: [Interrupting], yeah, it's really about future returns. Investors are obsessed with seeing down the line. It doesn't matter though. Pharma always gets its money.

Simon and Michael's exchange brought up the paradox of how pharmaceutical companies, with a pipeline of new products drying up and a tough regulatory environment, remain profitable. It also brought up the possibility of shareholder anxiety as a factor in Pharma's evolving position on partnering. Were shareholders concerned about running out of creative business strategies in lieu of real innovation?

During his presentation, Leonard outlined his partnering strategies:

> We will partner our internal talent with academics, and I think that this has been successful, but I think that what we felt we needed to do is to build a small biotech company within [company and specific division] and so we had to bring in people who could really understand biotech. Not to just understand new therapeutic concepts. ... We used to have lots of alliances, and they were very arm's length. I wanted to change this, and I feel strongly that we don't just sell a device, we sell a therapy, and we want a 50/50 sense with our deals.

As a strategy for inspiring entrepreneurial innovation within each company's own R&D divisions, building a small biotech company within biotechnology and pharmaceutical companies was trending. Given the limits of therapeutic understanding for building successful products, a necessarily collaborative approach brings together the expertise of scientists and those with expertise in the business of the life sciences. In this way, these social spaces could be viewed as contexts for a kind of hybridity, not merely bringing scientists into mere partnerships with existing company infrastructures. Instead, sociality was about fusing expertise together, folding it into new corporate formations.[2]

Significant venues for connecting were the cocktail hours and the catered lunch. At the St. Regis, lunch was held outside on an enormous terrace on a middle floor that afforded only brief views of the city on account of the sizable cream-colored tent in which about 20 tables were clustered. It was a working lunch, and a keynote speaker approached the podium and began discussing the "true cost of risk" for companies in the neuroscience sector. The speaker took up about a 30-minute portion of the lunch hour. In the sunny San Francisco day, a warm, muted glow from the sun lit the tent and produced a lush light.

The speaker, the CEO of a company that creates products to help improve the quality of clinical trials, spoke in a strangely grave tone about the impossible difficulties that pharmaceutical companies face in an overly burdensome regulatory environment. Despite his company's self-ascribed focus on increasing quality in clinical trials, his speech almost exclusively addressed all of the creative strategies that can be used to massage clinical-trial data and outcomes. The end of his presentation convinced me that his real goal was to attract new clients among the pharmaceutical, biotechnology, and medical device industry representatives in the audience. As he ended his short speech, he began to excoriate the risks and true costs of being innovative, finishing with dramatic stories about his experiences with large, expensive clinical trials and the role of awful government regulation in those costs. The speech became a harangue about regulation and the way that the larger "system" punished innovative pharmaceutical companies. A white-haired man to my left nodded aggressively in agreement. He then went on to bombard his rather disinterested neighbor with a peroration about taxes.

As the keynote speaker concluded, an orderly battalion of serving staff marched in unison into the room. The warm glow from the sun caused the metal serving platters to sparkle as they appeared and were deposited in front of each attendee. As Zack had intended, the end of the keynote left about an hour for lunch and mingling. Squinting at my name tag, the white-haired man to my left hunched down to get a better look, his chin approaching the height of the table. "So, Princeton, eh? What do you do there?" he asked, still squinting. "Well," I replied, "I'm a researcher. I'm interested in understanding more about TN." He nodded slowly as if to signal understanding and, clearly disinterested, returned to his filet mignon. Nothing tangible emerged in our interaction, and he soon revived his harangue about taxes with his now-inured neighbor.

To the right of me was a woman named Suzanne White. A young attorney, she gave me her card as a matter of custom. She explained that she works on issues of intellectual property, especially for life science and biotechnology companies. Based in Silicon Valley, her office was part of a larger international law firm that does a lot of work for "clients in this space." I asked her about her experiences working with entrepreneurs. She said, "Although I'm relatively new, I have had some experiences with university-based scientists looking to establish intellectual property protections." She continued:

> One problem that they now encounter is an aggressive intellectual property agreement on the part of universities, which lay claim to not only IP developed at the university, but also, to the extent to which they can, they try to lay claim to IP that they believe resulted from discrete research, learning or educational activities that happened at the university. Universities are really wising up to the revenue potential attached to their intellectual property and their resources. (Interview 11, May 21, 2010)

I asked if she worked with entrepreneurs as they formed relationships with companies. White replied that at her firm, those concerns were handled by a different practice. However, she did articulate a caution. Leaning in slightly, she said, "I know a lot of stuff about many of the companies in this room. *Honestly*, some of these guys will choke their own family for a good deal" (her emphasis). Then, as she sat back in her chair, she whispered with a chuckle, "Sharks." I asked what she meant. Eating her salad, she began gesturing with the other hand, "Well, you know, information is sometimes as valuable as IP, I've found. Remember, IP is only as valuable as the market."

At that moment, people began slowly exiting the tent, turning back into the building. As staff came out to collect the plates and trays, White and I went down toward the main hall. Around the corner from the registration table, there was a maze of hallways and informal areas. Up a half flight of stairs, there was an additional maze of nooks and informal seating spaces.

We found a quiet spot in a nook that looked out over the city. Peering out, she said, with a tone verging on shock, "It's a beautiful day!" As she laid her jacket over the side of the club chair, she continued,

> I mean that even IP is really only valuable in a context. I can't tell you how many times I hear about companies that purchase a small firm, but which write into the contracts these super long engagement or consulting times from the entrepreneur, sometimes for years just because of all of the contextual and domain information that he has and to which the IP is often irretrievably tied. In that way, these super bold IP agreements that universities are now requiring of faculty make sense. You really can argue that the university helped create the context where certain knowledge took place. Some entrepreneurs are now wising up and once they come up with an idea, suddenly [they] quit the university and wait a length of time before filing for a patent. Another example, [of the importance of context for IP], one that I'm thinking about that happened recently, was with a client who decided to sue an advisory firm that was helping him and his small biotech company. He had eventually fired that firm. He was an academic—a little on the crazy side. But he decided to sue them, claiming that his IP was valuable and only valuable because of a key relationship that the advisory firm had that was now lost because the advisory firm will no longer make the connections [since he had fired the firm and the firm controlled the ability of its advisors to interface with clients outside of the firm-client relationship]. He's saying that his IP is rendered valueless. This was something that I really learned in my time learning the ropes in this practice—that there is more to IP than what you might think. (Interview 11, May 21, 2010)

White's positionality made her explication especially useful. Her insight into the legal risks and entanglements for university-based entrepreneurs provided a novel angle into the relationship between scientist-entrepreneurs and their location in a complex social web of interdependence. Her own theorizing about the way that legal structures around IP impact collaboration and the inherently contextual nature of IP corresponded with my own hunch about the fact that TN's output is *rendered valuable* in and via given contexts—an issue that has been missing in modern analyses of science-technology distinctions and the problem of knowledge (Faulkner 1994). White eventually got up to return to the event. I stayed in the vicinity to look for a place to write.

As I peered down the half flight of stairs that led back to the main area, I saw Zack enthusiastically greet someone in the main area and quickly whisk them up the half flight of stairs past me to introduce that person to another. Zack was a crucial connector in this space, introducing advisory firms to potential biotechnology clients or connecting IP lawyers to startup entrepreneurs. TN could only occur via such social connections. White's story about the man who sued the advisory firm because the firm would no longer provide him with a key connection, which , he argued, subsequently rendered his IP without value made it clear that social networks constitute key aspects of IP value.[3]

The social is a crucial vehicle enabling what Kaushik Sunder Rajan and Sabina Leonelli (2013) outline as the *"trans-"* in translational science and research. In fact, White's example of the client whose IP could only "move" with the help of his relationships, as well as Zack's constant introductions, reinforced that the sociality I was witnessing at the Neurotechnology Investing and Partnering Conference, which constituted the very "mobility" that Sunder and Leonelli identify as integral to the concept of translational research. According to Sunder Rajan and Leonelli (2013, 466), "Mobility is therefore central to the idea (and ideal) of knowledge that animates translational research. In order to act toward the aspiration of 'improving human health,' biomedical claims, objects, and practices have necessarily to move across boundaries."

Still in the upper area, I sat down on a bench to record my observations. For university-based researchers who have made a discovery in their lab that they believe could be later translated into a useful therapeutic or diagnostic, finding access to capital and resources is essential. But on the path between discovery and invention lie expertise, pitfalls, and questions that the researchers must successfully navigate. These events function as a space where university scientists can introduce a laboratory discovery and render it—quite literally—meaningful and/or *valuable* in all of the diverse, subjective, and complex ways that translational value gets codified (Sunder Rajan and Leonelli 2013, 465). Here we see how finance helps to "matter" scientific findings.

The sequestered space in the upper area where I sat to write was contemporary and beautiful, with dark, muted rouge carpets that complemented the slightly glossy, salmon-colored, wallpaper. Momentarily deserted, the space acquired a Zen quality. In expected and unexpected hotel spaces, one consistently found art. There was a beautiful and intricate wire sculpture suspended from the ceiling over the stairs that connected the two spaces. Near the area where I sat and wrote, I noticed a cluster of framed images neatly arranged on the wall (see figure 5.7).

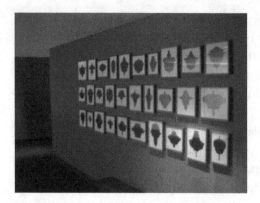

**Figure 5.7**
The Rorschach-inspired art installation at the St. Regis Hotel. The upper area adjacent to the event space with the Rorschach test–inspired art installation.

It drew my attention and I went over to get a closer look. As I approached, I discovered that the cluster of pictures consisted of dark, stylized images of inkblots, reminiscent of Rorschach tests, which psychologists used most widely during the 1960s in the United States. I took a picture of the arrangement. For me, the prominence of the Rorschach test was a result of its cultural appropriation in US television shows, where it was used during scenes that depicted interactions between a psychotherapist or psychoanalyst and his or her patient. Yet the Rorschach test was much more than this. Its inventor, Hermann Rorschach, a Swiss Freudian psychiatrist, had intended for it to be a tool that could be used for the diagnosis of schizophrenia (Rorschach 1921, 2008). Based on an experiment using 300 institutionalized patients alongside 100 control subjects, Rorschach created an inkblot-based diagnostic tool that used "Psychodiagnostic Plates" as a central element within a larger diagnostic method outlined in his 1921 book, *Psychodiagnostik*.

The serendipity is multiple. First, it is interesting that this early relic of psychiatric diagnosis had become relegated to an aesthetic role, a psychiatric and technoscientific anachronism, a cultural artifact now used within pop culture to gesture toward historicity and also, in this case, via its modern reinterpretation by an artist. Yet a second serendipity exists: this art installation is staged no more than 50 feet away from an event focused on, and in many ways charged with, collectively producing and predicting the future of psychiatric diagnosis. In some ways, it felt like a sort of museumification of tropes and modes of early psychiatric thinking and doing: a now-mummified artifact of a now-defunct biomedical presumption. It represented the impermanence of psychiatric understanding as well as the history of tools used in capturing and understanding human madness.

On returning to the main events, I plunged into another important strategy that marked event activity—partnering. By the time this conference happened in 2011, this model of partnering between Pharma and biotech firms was widespread among a variety of innovation-focused corporations across many sectors. The dialogue between Michael, the investor, and Simon, the Harvard scientist, about the firms staking out Harvard's bioscience faculty only served to confirm this shift further. The written description of a panel titled "Translational Partnering and Investing Opportunities" included the following summation articulating the relationship between academia and spaces of "opportunity":

> This session will feature selected cutting edge, research projects from academia, research institutes, and government labs ready for collaboration or hand-off to industry. Don't miss this showcase of the next generation of drugs, devices, and diagnostics selected from the top TN projects. Supported in part by a grant from NINDS, NIA, and NIMH.

After incorporating and filing patents lies the process of scaling: researchers must find investors to take their small, newly formed biotechnology firm to the next level. Often, they'll get an introduction to Zack. Michael Hall explained to me during intermission that Zack's introduction means a lot: "To have an idea vetted by someone who understands the industry and the space and the climate is really important. Certain introductions mean more than others."

While the importance of a social connection in the world of biotechnology is known among scholars who have explored the impacts of venture capital on the practices and the viability of biotechnology startups (Rabinow 1997; Sunder Rajan 2006; Shapin 2008; Sharp 2013), the reputational structures at work in investor introductions speak of an altogether different vector of meaning. Michael explained to me the delicate currencies accorded via crucial social introductions: "So, I definitely go and talk to scientists in their labs, but I pay attention when Zack says to me, 'Hey, I've got something here.'" Both the 2010 and 2011 Neurotech conferences included scheduled social hours where exchanges went from official to

**Figure 5.8**
At the Neurotechnology Investing and Partnering Conference, much of the program is devoted to providing scheduled time for networking. Additionally, the venues selected often include a panoply of rooms, lounges, and adjacent spaces where participants have discussions, entrepreneurs discuss their companies, and ostensibly, deals are made.

unofficial, and Zack continued to broker conversations and introductions (see figure 5.8).

As time went on, these social spaces became havens of trust. At the St. Regis, conversations would move from the area surrounding the main space down to the lounge areas. The shift from official to unofficial discussions was noticeable. I observed gradual changes in body language, sudden intimacies between two speakers. The intimacy of these conversations made them difficult to join, especially for a marked outsider.. All the while, Zack was not only arranging interactions and sharing among people; he was also consuming. He asked pointed questions while bringing attendees to meet others. He was constantly learning. It was because of this—his knowledge repository of interests, projects, and goals—that he was a consummate neuroscience broker. I initially thought of the Neurotechnology Investing and Partnering Conference, as a sort of marketplace. Indeed, the need for deal making—for investors, entrepreneurs, investment firms, advisory firms, lawyers—enabled it to be conceived in this way. This event, however, is more than a space for trading commodities.

Clifford Geertz, prompted in part by the theoretical poverty of rationalist models of economic behavior, which reduce all kinds of markets and marketplaces to the logics of competition (thus, rational choice theory), wrote about the role of information and knowledge in systems of exchange. Reflecting on his own research in the Moroccan bazaar in the town of Sefrou, Geertz elucidates the ineluctable functions and paradoxes at work around information in the bazaar economy. Writing about the ubiquity of ignorance inherent in the bazaar, Geertz (1978, 29) attributes the importance of information precisely to its tyranny:

> The search for information one lacks and the protection of information one has is the name of the game. Capital, skill, and industriousness play, along with luck and privilege, as important a role in the bazaar as they do in any economic system. They do so less by increasing efficiency or improving products than by securing for their possessor an advantaged place in an enormously complicated, poorly articulated, and extremely noisy communication network.

Thinking of the Neurotechnology Investing and Partnering Conference as a bazaar helps to make sense of its complicated sociality, and the way various stakeholders—investors, entrepreneurs, scientists, scientist-entrepreneurs, as well as biotechnology- and pharmaceutical-company executives—employ practices for modulating access and information. This contextualizes the edict of Leonard, the boisterous medical device-company executive who warned attendees to "be altruistic." This admonition, and perhaps desperation, reflected both his "privileged" position within the bazaar and his vulnerabilities from the poor articulations and extreme complications of the bazaar's "communication network."

The concept of the bazaar as an information economy also helps make sense of the information anxieties worked through during panel presentations. Panel titles simultaneously conveyed information anxiety and also promised information advantages: "New Investment Strategies for an Increasingly Unstable Neurotechnology Industry," "Emerging Treatments in Epilepsy and Pain," "Next Generation Psychiatry."

Geertz distills micromeanings that make information worthwhile. According to him, "Information search thus is the really advanced art in the bazaar, a matter upon which everything turns. The main energies of the bazaar are directed toward combing the bazaar for usable signs, clues as to how particular matters at the immediate moment specifically stand" (1978, 30). The notion of usable signs is quite apt here in that it specifies the process of reading indicators of potential—for example, recognizing a new startup working at the edge of the latest innovations in bioscience. Michael, for instance, is not simply in search of information. Rather, his constant conversations and his relatively small number of actual investments speak to this notion of combing—of looking for, as Geertz might put it, something like investable signs. One sees this ethos throughout the event. Notice how the description for the 2010 panel event "Investing in Neurotech" uses "usable signs" as a means to entice attendees to participate in the event in order to gain a competitive advantage:

### Investing in Neurotech Panel

A diverse panel of private, public and strategic investors will discuss their investment strategies. What does it take to get a neurotech company funded? What are the benefits and risks of business models in devices, pharmaceuticals, diagnostics and information technology? What are today's deal terms and valuations? What are the best exit strategies and how is neurotech comparing to other industries? (2011 Neurotechnology Panel Listing)

The notion of signaling usefully gets at the practices of the investors, in particular, who are not only looking for good information and strategic advantages, but must also *discover* opportunities. It points to the performative practices of those hoping to convey the right signs. Thus, Leonard's exhortation about altruism was also designed to create opportunities for discovery. This paradox—the need for an excess of information and exposure to an abundance of possibilities in tension with the search for rare, usable signals—produces a challenge. One needs a robust network and exceptional winnowing skills to find the needles in the haystacks. The "primary problem facing its participants is not balancing options but finding what they are" (Geertz 1978, 30).

Yet, amid the noise of the "bazaar," the polyphony of varied and situated voices, lay an internal structure of strategic intimacies. These intimacies are signaling processes between potential collaborators: Michael's leaning-in to impart information about the fact that "everyone's hungry" or Zack's strategic sequestration of potential collaborators away from the main event space. The abundance of intimate spaces and areas are designed to induce and stabilize momentary intimacies; what Geertz (1978, 32) calls "personal confrontation between intimate antagonists" is an essential element of the event's design. These moments can be thought of in terms of something like pollination. In the larger life cycle of TN, one could ask about the ways the entire process of translation depends on—and is perhaps sustained through—these arranged intimacies.

It's also worth noting the applicability of Geertz's notion of *clientaliza-tion*: a tendency for those involved in the bazaar to "establish continuing relationships with particular purveyors." Against any notion of pure depen-dency, Geertz's notion of clientship is adversarial, competitive, and egalitar-ian. Moreover, it is a means for the hazards and costs of information search to be reduced and systematized:

> Clientship is a reciprocal matter, and the butcher or wool seller is tied to his reg-ular customer in the same terms as he to them. By partitioning the bazaar crowd into those who are genuine candidates for his attention and those who are merely theoretically such, clientalization reduces search to manageable proportions and transforms a diffuse mob into a stable collection of familiar antagonists. The use of repetitive exchange between acquainted partners to limit the costs of search is a practical consequence of the overall institutional structure of the bazaar and an element within that structure. First, there is a high degree of spatial localization and "ethnic" specialization of trade in the bazaar, which simplifies the process of finding clients considerably and stabilizes its achievements. If one wants a kaftan or a mule pack made, one knows where, how, and for what sort of person to look. And, since individuals do not move easily from one line of work or one place to another, once you have found a particular bazaari in whom you have faith and who has faith in you, he is going to be there for awhile. One is not constantly faced with the necessity to seek out new clients. Search is made accumulative. (1978, 30–31)

In this view, clientalization produces a sort of stabilization, a means to manage labor divisions conceptually, and enables what could be called a kind of market analysis: a holding in place of knowledge, metaknowl-edge, and signs in order to ease processes of information and exchange. Consequently and concurrently, clientalization makes possible a neces-sary and systematized means of filling in for an inevitably impoverished system:

> Clientalization represents an actor-level attempt to counteract, and profit from, the system-level deficiencies of the bazaar as a communication network—its struc-tural intricacy and irregularity, the absence of certain sorts of signaling systems and the undeveloped state of others, and the imprecision, scattering, and uneven distribution of knowledge concerning economic matters of fact—by improving the richness and reliability of information carried over elementary links within it. (1978, 31)

Unlike the Sefrou bazaaris in Geertz's example, there is significant turn-over in the world of biotechnology and the corporate life sciences. However, given the increasing role of technologically mediated networks such as the professional social network website, LinkedIn, we see the ways that technology is positioned as a mediator in creating and sustaining intimacies. Given these technologies' capacities to organize information about divisions and competencies, digital mediation has become a significant tool of clientalization complementing the on-the-ground social process. Michael Hall, for example, quickly brought up LinkedIn to answer a question about the social connections of those attending the conference. White, the Silicon Valley IP lawyer I met at lunch, still needed to attend these events in person to create the sorts of intimacies—later codified via LinkedIn or initially opened up through LinkedIn—that were optimally formed by sitting around a lunch table in the middle of a sunny day in May handing out business cards to prospective clients.

Geertz's articulation of the relationship between structure and knowledge in the bazaar functions as a complement for—if not an antecedent to—the material-semiotic analysis of material and social worlds conceptualized in theorizations such as those within actor network approaches (Law 2009) and that of the trading zone in Peter Galison's work on the many cultures of microphysics (Galison 1999).

The social problems of knowledge and information for which processes such as clientalization constitute a solution are conceptually useful for thinking about the thoroughly social means through which the "gaps" at work in TN are traversed. If the continuum between a scientific laboratory and its commercialization includes a meeting of materials and epistemologies—animal models and fertilization tools, glial cells and psychopharmacology—social arrangements can be seen as levers and bridges that fill in the gaps over which translation must occur. In this way, the very architecture of the St. Regis Hotel seemed to work in concert with the event's necessary social flows and designated meeting areas. The social networks and LinkedIn connections alongside the lounge space and mezzanine levels constitute a comprehensive ecosystem. These socially enabled cross-pollinations enable the essential bridging activities of translation. It is via these social and material bridging systems that one can locate not only the categories of knowledge mobilities engendered and required under the regimes of translational research (Sunder Rajan and Leonelli 2013), but also knowledge in transit moving in particular and specific *directions*.

## 5.5   The Subjects and Objects of Neurotechnology Markets

The sociality of TN is also useful in tracing the avenues through which pharmaceutical presumptions may impact upstream practices in university-based science laboratories. For a university scientist attending the Neuro-technology Investing and Partnering Conference event, experiences with investors and articulations of investor value can influence the decision to select one scientific question over another. One might ask what the impacts are of the endless panels of scientists and entrepreneurs (funded and unfunded) in discussions of fruitful and nonfruitful trends in life science research on the micropractices of university scientists.

Scholars such as economic historian Philip Mirowski (2011) have traced the shifts in university practices under neoliberalism, yet these structural and institutional evolutions are not the same as the ethos, the long-harangued scientific sensibility thought to animate and undergird science under classic, triumphalist accounts of scientists and their vocation (Haraway 1997). Efforts of scholars to trace trends toward commercialization in the university itself have been significant, and yet the imprint of market forces on the work of scientists in terms of ethos is less clear (Gieryn 2008; Hong and Walsh 2009; Kleinman 2003; Krimsky 2004).

However, measuring the impact of market forces on the practices of individual scientists, their decision making, epistemologies, and subjectivities, would constitute a Sisyphean task. In the absence of data about a scientific consciousness, Mirowski (2011), in the middle of his historiography, inserts a fictional scientist, Viridiana Jones, whose subjectivity becomes a space in which he theorizes relationships between the encroaching commercialization of research and basic science in individual scientific practice and thinking.

For historian Steven Shapin (2008, 251–252), market forces have not entirely displaced trust and virtue as the everyday currency of scientific collaboration and work even in the contexts and soft erosions of late modernity's "entrepreneurial science." Yet if, as scholars have ardently suggested, there is an indubitable shift toward increasingly market-oriented (Berman 2012), patentable (Mowery et al. 2004), and commercializable (Bok 2003; Lowen 1997; Shapin 2008) academic bioscience research, it is legitimate to ask whether the need for venture capital and the language of translatability necessarily prefigure research agendas at the modern research university. Does an assumption of a psychopharmaceutical endpoint necessarily produce psychopharmaceutical thinking? Are there risks from this mode of thinking as it relates to other kinds of marginalized data, questions, and possibilities? During one panel focused on collaborations between large life science companies and universities, a partner at a New York–based venture capital firm said to the audience of university scientists, investors, and others, "The goal is to think about how Big Pharma and Medtech are thinking about neuroscience."

A first reading of this statement concerns the aligning of smaller firms' strategic thinking about neuroscience research with that of larger pharmaceutical and biotechnology companies. However, there is a potential second reading of his statement, which would be about scientific thinking as it impacts on-the-ground laboratory practices. In other words, if "the goal is to think about how Big Pharma and Medtech are thinking about neuroscience," this suggests that universities and scientists that hope to one day partner with pharmaceutical and device companies (or to create knowledge considered valuable in that realm) may begin the process through modulating their on-the-ground laboratory practices and research agendas in order to complement "how Big Pharma and Medtech are thinking about neuroscience." Here, "thinking" becomes the first step toward legibility.

Beyond the increasing commercialization of academic bioscience lies the question of the implications for biological meaning. How does one distinguish between biological observation and a finding or distinguish a finding from an invention, even if that invention is merely a new means or method for revealing biological material? Underneath concerns about permuted scientific practice emanate questions about the changing semiotics of biological material under neoliberalism.

## 5.6  "The Secret of Weight Loss Is the Scale"

Scholars have shown just how lucrative psychiatric disorders and global biopsychiatry, in general, have been for pharmaceutical companies *despite known problems* (Applbaum 2009b; Ecks 2010; FitzGerald 2005; Gagnon and Lexchin 2008; Healy 2012). According to the *Neurotechnology Industry Report*, in 2008 neurotechnology companies generated revenues of $144.5 billion, representing 9.0 percent growth.

Within this paradox, TN can be seen not only in terms of its epistemic and financial risks, but also in terms of its capacity to mitigate the risks that had ballooned based on a history of problematic practices by pharmaceutical companies. Pharmaceutical executives at the Neurotechnology Investing and Partnering Conference, events displayed a relative agnosticism regarding the absolute necessity of deep neuroscientific knowledge, and Simon spoke of an emerging industry preference for neurotechnologies (by which he really means devices and diagnostics, rather than psychopharmaceuticals, though these are also technically neurotechnologies) over pharmaceuticals because of their lower risk profile. In other words, neurotech devices have a risk-mitigation function for TN.

There is another instance where risk is prominent. One panel moderator posed a familiar problem: "Boston Scientific is selling its neurostimulation division. Does this reflect waning interest?" Boston Scientific was one of several companies that either sold or shut down their neuroscience research groups in 2010. One investor, who directs his own eponymous venture capital firm, said, "We like ... diagnostics." He said that his firm sees it as a huge opportunity and is actively trying to figure it out. Of course, what is intriguing is that his statement, "We like diagnostics," was uttered in a way that conveyed the sense that diagnostics are *something* or at least *one thing* that his firm embraced. In response to this, Simon interjected: "One of the challenges in diagnostics, at least the way that we use them now, is that there is an interplay between diagnostic and therapy. Some simply molecular diagnostic or some prognostic market is a huge market going forward." In other words, products that merely diagnose or that offer predictions are important market opportunities that involve products that aren't technically cures and which have large market sizes.

Simon's articulation is instructive. But it is also important to note how the move toward diagnostics reduces risk. Diagnostics themselves are less risky than therapeutics. Moreover, Simon rightly acknowledges that a problem for diagnostics can emerge when tied to an existing therapeutic solution. However, diagnostic tools that are untethered as a "solution" effectively commercialize a patient's own biology. For example, molecular therapeutics that rely on genomic science simply offer predictions about tolerances. They are less tethered to actual therapeutics. In fact, molecular therapeutics make *biological information* their primary product (Fortun 2008; Lippman 1992; Savard 2013). In other words, a molecular diagnostic product monetizes sheer biological material alongside an attached clinical interpretation of its meaning. The logic of this strategy became clear during a morning panel at the 2011 event titled "Frontiers in Neurotechnology."

Laura was the CEO of a startup, SkyBridge Diagnostics. SkyBridge Diagnostics is an early-stage company focused on the development of blood-based tests for neuropsychiatric disorders. "In vitro diagnostics," she called it. Laura began her presentation with unmitigated scientific confidence. "We take the *biological approach*," she said, quickly scanning the audience. She then delved into an explanation about her company's current project and its partnership with a well-known hospital in New England: "We're in a network with [a prestigious New England hospital] to develop a blood test for bipolar disorder. Why do you need this? When you examine the challenges experienced at the clinic, it's hard to argue against additional biological tools or some additional diagnostic tools."

Laura revealed several important interplays at work in terms of clinical decision making. First, we see this aphorism that *more* science simply can't hurt. In this view, additional tools contain value in and of themselves. She promoted the prospect of a neurotechnology such as a blood-based neuropsychiatric diagnostic that produces mere information in a context in which mere information can be considered clinically valuable or, at a minimum, bereft of any real harm. However, her aside that "it's hard to argue against additional biological tools or some additional diagnostic tools" in the context of the clinic corresponds with the information hegemonies that Dumit articulates in his discussion in *Drugs for Life* of the complications that build up around the problems of screening, especially as it is distinguished from the thresholds of diagnosis. Recapitulating internalist articulations from the researchers and designers, Dumit (2012a, 146) recounts several questions that emerged about the purpose of screening tests for depression:

> Was it to indicate referral to a psychiatrist? Did the screen itself constitute diagnosis of depression? Was there any evidence that screening improved patient outcomes? ... Choosing to screen and setting a threshold thus involved a careful assessment of the entire healthcare system and the effect that the screen would have on each part of it. In the language of science studies, screening was understood to be part of a large-scale sociotechnical system. ... Screening also raised the socio-medical problem of confusing the screen with the diagnosis and with proof of illness. Would a screen given to primary doctors to pass along possible depressives to psychiatrists to be properly evaluated come to be taken by the

primary care doctor and patient as a diagnosis of depression and indication for treatment? These problems were ... being raised by the same researchers who were conducting the surveys, studies, and clinical trials, and designing new and better screening tools.

Laura's confirmation of the accretive nature of diagnosis and treatment at work in the clinical encounter represented the information itself as beneficial for patients. However, as the concerns in Dumit's account illustrate, the problem is one of ontological conflations and confusion: Is screening the same as a diagnosis? Is a diagnosis the same as proof of an illness? How might various diagnostic test results actualize or eliminate the reality or unreality of a disease? In this sense, the screen becomes a space in which to see various forms of material and semiotic actualizations, actualizations that also produce confusions and ambiguities that are acknowledged as such, but that are still understood as not constituting patient harm. Hence, Laura's easy, nearly axiomatic apothegm, "it's hard to argue against. ..."

The question of whether the *diagnostic* she discusses in her scenario would transform into a *screen* garnered greater importance as she continued talking:

> ... and so there is a diagnostic continuum that exists, half of the reason, maybe, that contributes to the failure rate of the initial use of drugs. So the efficacy of the drugs is compared against high levels of clinical accuracy and so the blood test— we're talking about a blood test—that approaches that level [a clinical research level] of high diagnostic accuracy and taking it to the clinic. So, with that, we hope to [be] able to see that these patients have better control of their disorders. Patients are seeking blood tests. We believe and the psych community and the primary care docs agree that it helps for the patient to see their disorder. It helps patients to understand—for patients to understand that it's ... that it's biologically based.

Laura's scenario includes primary-care physicians, a convenient obligatory stage (Callon 1986) within the sociotechnical system of the blood test, rather than placing these tests exclusively within psychiatry. However, equally interesting was her suggestion that the higher level of diagnostic accuracy provided by her blood-based diagnostic tool might offer a solution to problems of psychopharmaceutical efficacy.

While it ushers in the promises of personalized medicine, biological material here becomes a resource for problems of already existing therapeutics. I noticed similar repurposing of patient biology during a panel on neuromodulation devices. One presenter advocated for a new way of thinking of brain devices as a means by which implantable biotechnologies might become a source for better medical treatment. According to that presenter: "Don't think of the little implant as a little implant. Think of it as a computer. There will be genomic and feedback info based on this. We need to make sure that healthcare is a zero sum game. We have to make sure that medical practice changes so that we get perfect treatment."

In this mode, one sees implanted devices as a meeting between, in his words, "informational, pharmaceutical, and device-specific approaches." However, these approaches all work in tandem with a patient's own biology—as the source of genomic feedback, for example. One's biology becomes a part of the therapy just as it also becomes the laboratory where data, modulation, and reporting all come together.

Accordingly, Laura's explanation about the *utility* of blood testing makes clear how a blood-based test not only offers a means of making an informational commodity out of blood, but becomes a means for deeper entanglements with pharmaceuticalization and the clinic. While the promises of personalized medicine are based on the possibilities of more tailored and more efficacious medicines, the logic enables a *systematic undoing* of the barriers to pharmaceuticalization, thereby creating a need for greater access to patient data and more sharing. After her statement, Laura then indicated that the *real* benefit is in the fact that tools like the blood test create "conversations between patients and their doctors."

According to her, "If just 10 percent of overall conversations could lead to a blood test, it leads us to a huge market potential." In this way, one sees that sheer information is in no way neutral. Rather, it has many implications for ontologies, confusions, risks, and therapeutic and economic possibilities. We also see in Laura's numerical and market reasoning ("just 10 percent") the kinds of power accorded to numbers and market sizes in the clinical encounter, a phenomenon that Lochlann Jain (2013) chronicles in her research on cancer and the effects of numerically driven diagnostic and clinical cultures. Yet the most evocative image during Laura's statement lay in her remark about the importance of helping patients "see" their disorder. This conflates biological material—blood and all that is contained within it—with patient "disorders." Her statement inspires both visions and questions: What does it mean for a patient to "see their disorder"? How might TN's leveraging of sheer biological material work in relation to the construction of psychopharmaceutical subjectivity, or what Janis Jenkins (2011, 66) has unearthed as the shaping of "the pharmaceutical self"?

Yet the claim also conflates a diagnostic and its disorder. This produces questions about the mechanics of diagnostic realization. Through which practices does a diagnostic test transform into visions of one's medical disorder? Might this moment of emergent social and material practice at work between the clinician and the patient represent something like psychopharmaceutical intersubjectivity? This question suggests the discursive production around biological material, which is in itself an ineluctable part of the "product" of diagnostics. In other words, the diagnostic test is inextricably fused to the discursive authentications and scientific explanations of the clinician. However, an additional and critically important part of the therapeutic role of the diagnostics was elegantly summed up in Laura's statement about biology: "It helps patients to understand—for patients to understand that it's ... that it's biologically based."

This statement is telling. Laura's initial construction, "It helps patients to understand," and her midsentence clarification, "[it helps] for patients to understand," both align with the objective and subjective utility of biological meaning making in the realm of the clinical encounter. In the former, the diagnostic helps patients to see and understand their own disorders. In the latter, Laura affirms the benefit to larger structures—the clinical encounter, patient education paradigms, pharmaceutical practices—of having patients achieve psychopharmaceutical realization. In this sense, the critically important part of the therapeutic role of diagnostics is the capacity for a patient's biological material to effectively conjure up epiphany and recognition. Perhaps these are the material phenomenology and associated technological accompaniments that enable something like an Althusserian interpellation—that is, the material means through which, according to Althusser (1971, 174), "ideology 'acts' or 'functions' in such a way that it 'recruits' subjects among the individuals (it recruits them all) or 'transforms the individuals into subjects.'" Interpellation becomes not only the material means through which an ideology is effectively activated within and transferred to given subject, but also a means of realization that works through the subject's "recognition" of the self.

The importance of this productivity around biological material becomes clearer as the conference goes on. The movement of the industry toward genetics enables and constitutes a solution for the problem of the messiness of psychopharmacological presuppositions and the fact that psychophar-macological models simply don't adhere to the same kinds of science and efficacy requirements as other drugs (Applbaum 2006, 107). By this I mean that these diagnostic-based strategies enable a repurposing of biological material—in this case, via genetics—as a solution for the problems of exist-ing scientific models and patient care. In a context in which "you have no models," this idea that one's own body can become the means and the evidence for intervention is powerful. The productivity of using one's own biology as a product in concert with diagnostics was elegantly summed up in a terse adage stated by a young CEO of a company that sold at-home sleep apnea diagnostic tools: "People can use their own data and to try to become more healthy. Remember, the secret of weight loss is the scale."

## 5.7   Discourses for Biological Material

Linda is a neuroscientist from Colorado and a tenured professor. And like many other scientists at the Neurotechnology Investing and Part-nering Conference, she is also an entrepreneur, looking for both funding and important social connections in order to move her small biotechnol-ogy company to the next level. At her turn, she approached the podium. Middle-aged, she wears navy slacks and peers through enormous glasses. As her PowerPoint slides are loaded onto a large screen, she begins her pre-sentation, which varies minimally in structure from that of all of the other startups at this event and most others that I attended over several years. These talks—the startup investor pitch—further validate these spaces as neuroscience pageants, or perhaps more precisely, bazaars for or theaters of epistemologies (see figure 5.9). With vivid PowerPoint images of bright brain scans and boisterous visions of biotechnology futures, the event is both a science carnival and an elegant marketplace for disease.

Like clockwork, Linda introduces herself and her findings: "Hello. My name is Linda Roberts. We have identified a novel mechanism of action and a new class of therapy in a large marketplace where existing mechanism knowledge has led to unmet need." She continues through her presentation, being sure to provide data about the large market that her product could serve. In a strange moral-linguistic ornamentation, large markets are often spoken of with the phrase "*unmet need.*" A large "unmet need" or poten-tial market is an imperative for investors, and thus the phrase permeates

**Figure 5.9**
Image of a presentation at the 2011 Neurotechnology Investing and Partnering Conference, held at the St. Regis Hotel in San Francisco. This presenter at the podium is an entrepreneur/scientist introducing his company and its novel findings to the audience.

startup presentations, written material, and private conversations. In these spaces, well-done science without a market is almost meaningless. Her phrase "where existing mechanism knowledge has led to unmet need" is a euphemistic way of saying that the current product is clinically ineffective. In the end, as is obligatory, she concludes with what is colloquially termed "the ask." Somewhat timidly she announces, "We need $7 million." The audience barely blinks. Flurries of startups ask for an assortment of million-dollar amounts. Some in the audience simply write down notes during the pitches. Occasionally, the startup entrepreneur is approached at the end of the pitch only to be taken to the hallway or a nearby lounge area, purposefully designed to encourage informal interaction.

There is interest in Linda's startup. Her company produces a gene-based therapeutic to treat neuropathic pain that works by stimulating the body's anti-inflammatory cytokine. Given the molecular basis of her company's product, her work is timely in the context of commercial neuroscience's move toward molecular biology and gene-based research. In response to a question by an audience member, she outlined several challenges she experienced in her initial search for the right compound to constitute a safe therapeutic:

How do you identify the numerous compounds that are effective, and which are reasonable candidates to try to take to clinical trials? There are excellent research compounds ... but maybe that compound's mechanism of action is that it's a metabolic poison, so there are a lot of examples from the research end of it, but in reality, it [a compound that acts as a metabolic poison] couldn't possibly go to humans. ... Targeting: how do you get a compound to work on the glial receptor? Targeting is a problem in getting it to the site that you want. ... Glial cells are immune-like, and so changes to glial cells will have that consequence as well.

Though Linda's statement here usefully explicates challenges involved in the molecular search for productive compounds, this statement and the ensuing dialogue are also acts of biological *problem solving*. Here, the problem includes the vicissitudes of biological inadherence, made clear in her example about the immunological "behavior" of glial cells, and about biological (in)compatibility between human parts and that of animal models. This was not merely an articulation of her own research journey. Linda's presentation indexes an eventwide dialogue about making sense of complicated and unruly biological material. How does one deal with mechanisms of pharmacological action? Is knowledge of such mechanisms a necessary component of the research and design process? How does one deal with the unruly biologies of engineered mouse models? Linda's statement about the immunological character of glial cells also maps onto another, trenchant biological problem of adherence and efficacy between therapies and patients' individual biologies. Thus, the turn to genomics is not only about the failures of premature therapy, or the riskiness in the too-complicated biology of mental illness; it's also a means to use biology to understand pharmacogenomics in ways that can lead to personalized medicines (Ninnemann 2012). In this way, pharmacogenomics becomes a means to deal with problems of biological variability optimistically assumed to be at the heart of poor psychopharmaceutical efficacy.

Thus, discussing the methodological problem of glial cells functioned as a sort of collective walking-through with the audience of the various cellular and compound possibilities and their respective limitations. One locates in Linda's extrapolation an epistemological journey around compound possibilities and routes, known cellular facts and unknown ones, productive questions versus nonproductive ones, as well as a kind of collective reflection about potential paths and pitfalls.

An additional instance of problem solving during the conference focused on biological material. During a panel on next-generation psychiatry, Luis Manuel, a scientist-in-residence at a life science venture capital firm, just presented a list of failed or discontinued compounds that comprise existing knowledge in the search for novel drugs and devices in psychiatry. His PowerPoint presentation consisted of compounds and what he called "approaches":

Compounds:

1.  Beta-3 Agonist

2.  NK@

3.  Etc.

New and Upcoming Approaches:

1.  Neurogenesis

2.  Combos

3.  Circadian genes

4.  miRNA

5.  Epigenetics

6.  Optogenetics

7.  Ultrasound

8.  Biomarkers (years away)

9.  Diagnostic devices

10. BBB penetration

11. iPS cells

Luis only minimally accompanied his presentation of compounds and approaches with dialogue. For example, when he got to circadian genes, he mentioned more "focus there in research for psychiatric disorders" and ended there. When talking about miRNA, he explained how its real value is in its use for blood-based diagnostics. For biomarkers, he simply said, "We are years away." There was no additional discussion, as if the objects in the presentation spoke for themselves. The ensuing silences during his presentation did not cause audience confusion or unrest. Attendees on both sides of me continued noting, writing, and checking their BlackBerry devices. These absences of explanation and discussion showed the importance of these biological materials—and compounds in particular—as objects around which to find solutions, begin investigations, and produce narratives.

Contexts such as these illuminate the space of the Neurotechnology & Industry Partnering Conference, as well as TN in general, as means for creating not only solutions from the problems of biological material, but also potential market solutions for existing or potential biological *disorders*—in all of their social and technical challenges. In the partnering of biology and disease categories one sees how TN produces what scholars have called *patients in waiting* (Dumit and Greenslit 2005) linked to futures and markets. At one the Neurotechnology & Industry Partnering Conference panel, Harry Thiel, head of neuroscience for a major pharmaceutical company, offered an overview of the role of *connection* in the various stages between "candidate selection" and "proof of concept":

> We begin by looking at proof of mechanism. Investment decisions are approached. Most clinical trials fail because [the underlying facts] don't work [because the science is bad]. This is in contrast to the me-too companies which try to get your data to work [through data manipulation or other strategies]. We know the way that functional neuroanatomy works because we know where we should be—which is to say that we know the science well enough that we know if your little experiment is based in fact or not. Linking disease to understanding and clinical outcomes is the first step, and then we try to link it to the right patients; the right drugs to the right patients.

Harry's biography charts his background at two biotechnology companies and at many of the largest global pharmaceutical companies where he held leadership roles. Among all of the speakers at that event, he was perhaps the most experienced and most senior neuroscience-focused pharmaceutical executive who had an impact on the direction of neuroscience within Big Pharma. He may also have been the person closest to the "realities" at work in commercial neuroscience. I tried to follow up with Thiel but was never able to reach him even via email. At the end of his presentation, he thanked the audience: "Thank you for sticking with neuroscience and as business models change," he said. He quickly got into the back of a town car and was whisked away.

Harry's jagged refrain, "your little experiment," suggests that his larger statement was directed at the small firms and university academics with whom his company may potentially partner. His tone was not one of collaboration; it reeked of unrequited hierarchy. Unlike the pharmaceutical-company executives at the Boston event who displayed relative agnosticism about MoA, neuroanatomical facts mattered according to Harry. This would suggest that he does not treat brain science as a pool from which to simply cull narratives and explanations. For him, the brain's materiality consists of a sort of functional grammar like one finds in foundational psycholinguistics. His aside, "we know the [neuroanatomical] science well enough that we know if your little experiment is based in fact or not," is a statement about materiality, presuming that experiments that draw on or interact with the *right kinds of brain areas*—those known to correspond with disease states—get coded as more likely to be based in "fact." In this sense, there is a literal matching process between the biological material of the brain and "disease," at least in Harry's scientific estimation.

In his next statement, "Linking disease to understanding and clinical outcomes is the first step," Harry uses the notion of *linking* apparently to reference a process of proving the relationship between a disease etiology and known scientific facts about the brain and then using that knowledge to understand a potential, therapeutic clinical effect. He invokes *linking* again to discuss the matching of interventions to the right patients. This latter point and his statement about retrospective analysis both refer to the pharmacological importance of understanding individual biologies, diagnoses, and, importantly, tolerability.

However, in a more constructivist sense, Harry's use of the notion of *linking* makes visible the epistemic practices that allow neuroscience facts to be *connected* and *or tethered* to particular brain areas, and for that *connection* to confer the status of fact or likelihood of fact onto a hypothesis. Similarly, his statement about "linking disease to [scientific/ biological] understanding" underscores the discursive mechanics involved in suturing biological understanding to diseases. Yet for Harry, good translational work is not merely translational; it is, in his words, *transitive*. Rather than a process that relies on a kind of *biological duress*, he advocates a sort of biological "listening," which allows for something like an autobiological translation. During the panel he said, "Transitive, not only translational medicine. It tells us what the next set of studies should be. So, again, our approach right now is complex but we're really trying to match the right patient for the right outcomes, as a clinical development tool. Let the biology dictate what we do rather than force biology to fit a particular need."

Here we see additional evidence of the way the conference operates as a means for collective biological problem solving. Harry's admonition about allowing biology to "dictate" scientific investigation, rather than the inverse, speaks to an epistemological proposition. It also, perhaps inadvertently, implies that *forcing science* through creative means such as clinical-trial and data-contouring practices, or hiding internal clinical-trial data, marked the history of Pharma's practices regarding pharmaceutical research and development.

What did he mean by *transitive*? The notion that translational medicine merely provides *markers* toward further research suggests, in fact, basic research. Transitivity may be a discursive means of allowing a dearth of new therapeutic results from TN to still be understood as a *product*. In my view, merely creating narratives and affixing them to biological facts becomes part and product of the translational enterprise even where there is no immediate, actualizable intervention. In his discussion of the many mobilizations that molecularization enables under what he terms "molecular biopolitics," sociologist Nikolas Rose (2007, 15) writes of molecular biology's transitivity, and especially its capacity for "disembedding":

> Molecularization strips tissues, proteins, molecules, and drugs of their specific affinities—to a disease, to an organ, to an individual, to a species—and enables them to be regarded, in many respects, as manipulable and transferable elements or units, which can be delocalized—moved from place to place, from organism to organism, from disease to disease, from person to person. ... Molecular biopolitics now concerns all the ways in which such molecular elements of life may be mobilized, controlled, and accorded properties and combined into processes that previously did not exist.

Scholars such as Sarah Franklin (2003), Margaret Lock (2002), and Rose (2007) have created a body of scholarship focused on how to make sense of contemporary bioscience, its emergent subjectivities and capacities, and in particular, the ways that emerging bioscientific capacities bring into question long-held narratives about life (Cooper 2008; Sunder Rajan 2006) and death. Much of this work (Helmreich and Labruto 2018) takes biopolitics as a key axis through which to investigate contemporary concerns.

However, for all of the mobilities that molecularization and translational research both enable and compel, a question still emerges about how directions for research are determined. The industry's channeling of goals for TN, the determination of directions for knowledge in transition, and the application of directives (literally) for scholarly agendas toward some particularities rather than others, show directionality. Beyond indeterminateness for biological material, a study of the contemporary capacities of bioscience and its subsequent transitivity invites scholarly work mapping not only these *directions*, but also the obfuscations and partialities that linger from foreclosed histories.

Thus, a concept such as molecularization is useful because it produces a way to make sense of the capacity for biological materials to be rearranged. However, there is a need to understand how these rearrangements engender certain outcomes over others. There is a need to understand phenomena such as molecularization as an exercise in the ontologies of partiality. The very capacity to separate and rearrange glial cells for experimentation indicates not only mobility, but, to the extent to which the glial cell is still semantically tied to the human, to a mouse model, or both, it constitutes the production of part of a *subject*. Against simplified narratives of reductionism or flattening, an ontology of partiality looks more like pixelation. Thinking with the parts of a subject emphasizes the question of bioscience's connections to the individual. In this sense, the slogan attached to the small silver brain pin I received at the beginning of the 2010 conference, *"Give the brain a voice,"* felt strangely literal. It was also a directive: one that animated this hermeneutic community of practice.

Indeed at the Neurotechnology Investing and Partnering Conference, partnering was a birthplace for all kinds of coupling, between a disorder and explanation, between patient explanations and biological knowledge, between biological knowledge and emerging diagnostics, between university laboratories and corporate laboratories, between pharmaceutical companies and biotechnology startups, and perhaps as a space for connecting biological materials with other biological materials. In this way, this event functions as an essential laboratory for TN.

## 5.8   Subjects Partially Constituted or Ethics for a Partial Subject

Near the end of the first panel, focused on the current state of investments in neurotechnology, the moderator asked panel members about their areas of interest. The panelists offered an array of answers:

"Neurodegeneration, pain, and epilepsy."
"I'd echo that, and we like the neurogenesis area," another investor concurred.
"Pain, neurodegeneration, schizophrenia," another offered.
"Severe disorders; … things that are generating some 'exclusivity' to make it more of a challenge to genericize," yet another investor said.

"It's not the most ethical, but we'll take a drug out there and increase the price, even if it's an orphan medication," someone indicated, referencing the pharmaceuticals and biotechnologies developed for rare diseases. What the investor likely meant by "not the most ethical" is the fact that patients who have no other therapeutic options are forced to pay higher prices for these drugs, which, in effect, provides pricing protection against pharmacological risk that is somewhat similar to that of a patent. The moderator enthusiastically agreed: "There is a lot of money to be made there." The sense of the *severity* of a disorder constituting a kind of market, an opportunity, invokes a complicated moral juxtaposition. It forces a convoluted thinking exercise: How does one monetize the reality of a patient's epilepsy? Can one excavate an opportunity hidden within a family's piercing desperation?

There is something else that ties all of these investor preferences together. The investor willingness to exploit "rarity" in the form of rare diseases and to go after "exclusivity," in the form of protection strategies such as those afforded by intellectual property, show two cases where State exception becomes the vehicle for corporate strategy aimed at market dominance. That both forms of exception (rarity and exclusivity) can be useful profit strategies creates questions about ways that state-sanctioned exceptions can function as variegated entry points for the expansion of capitalist markets.

And yet patients still exist in these formulations. Visualizations about severity and rarity—the investors' quick listing of trendy, interesting disease/biological states such as "neurodegeneration"—all implicate subjects. I had to rethink my assumptions that patients were essentially missing during these discussions. Patients were indeed "in the room" on that gray May day in Boston. However, I began to understand these subjects as only *partially constituted*, or perhaps more precisely, constituted in and via (their) *parts*.

For example, much of the May 2010 conference focused on finding new pathways to new markets and new verticals. Vertical markets here reference the attempt to use existing products for novel potential customers, the use of existing technologies toward new products, and expansion of opportunities in the form of global markets. During the part of the event that put a spotlight on new startups, Roger, founder of a neurosoftware startup, pitched his company to the audience of investors and researchers. According to him, he had created software that would provide "real-time" fMRI-based therapeutics and diagnosis. His software would obviate the problems of patients. He used the example of an everyday encounter between a patient and a doctor and the problem of getting to a patient diagnosis: "It's upsetting that [if] you want to get a diagnosis, you have to ask them how they *feel* [his emphasis]. This is a shame and with this technology you will see mechanistically based, quantitatively based information. You will be able to diagnose real-time."

Several things stand out in this particular diagnostic vision. Certainly, in one view, the marginalization of experience enabled within Roger's clinical scenario—that of the patient as well as the intersubjectivity between the patient and the clinician—constitutes a kind of erasure or reduction. Nevertheless, I also want to suggest a different view. Perhaps it is also about the subject constituted in his or her neurological *parts*, as partially constituted: a partiality made manifest through the marginalization of the patient's own phenomenological representations.

Another example of the sort of diagnostic vision that Roger articulated came during the presentation at a different Neurotechnology & Industry Partnering Conference by Harry Thiel, the prominent head of neuroscience at a major pharmaceutical company mentioned earlier. During a preview of his division's current projects, he explained how his company was working on a diagnostic tool in cooperation with General Electric's research support that would enable earlier detection of dementia using voice-based diagnosis:

> Physicians will be interested in diagnosing this early, and we don't want to wait until dementia [appears], but people aren't tracking themselves. So, we're looking hard with GE to cull all of the databases together to develop noninvasive tests that could be much more widely used. Here is something we're really interested in, voice-based diagnoses. Patients as they progress or as physicians identify more and more risk, we move them into more invasive [tests]. Let's say a spinal tap.

Here we see a similar kind of confusion between diagnosis and screening as well as a desire to create "noninvasive tests that could be much more widely used." Voice-based dementia diagnosis had been an idea in circulation in the neurotechnology community for a few years. The possibility of such a tool is based on research that suggests that dementia and Alzheimer's disease indicators can often include a discernible change in speech patterns (Baldas et al. 2011).

Nonetheless, as was the case with Roger's scenario, the diagnostic formulations presented here render patient subjectivity as problematic, a hindrance. The notion that "people aren't tracking themselves" becomes a means to look for dementia as it is manifested within the *parts* of patient subjects—which, in a rather grand irony, is that of the *voice* of the patient. A voice-based diagnostic test would work by using a natural language processing tool that captures a patient's voice and compares it against itself at an earlier time to measure any changes. The splicing of the patient's voice from the actual patient as a means of diagnosis also creates questions about the relegation of the "embodied" voice in these diagnostic regimes.

In one view, this diagnostic future, in its various technical, bodily, and digital mimicries, was reminiscent of the repetitions at work in the many images in the inkblot exhibit, although the resonances extend beyond mere aesthetics. For while Rorschach sought to find truth using image-based prompts and the voice-based diagnostic seeks to employ language-based prompts, in both cases, mental disorder is assumed to be locatable in the inevitable betrayals of a diseased person's speech.

Nevertheless, in the diagnostic visions—both that of the voice-based diagnostic for dementia and Roger's real-time neurosoftware solution—technological traces are codified as representing brain truths, which come to represent the brain *speaking*. In the diagnostic vision for the voice-based dementia diagnostic, the voice of the subject becomes the voice of and for the brain; the translational vision relies on an epistemology of parts.

Consider another example of what I'm calling an "epistemology of parts." While I pointed to the voice-based diagnostic technology to demonstrate how subjects can be constituted via their parts, one also sees this constitution of neurotechnology's subjects even in the question of interchangeability between animal models and humans. Parts become the key to an intellectual exercise in which animal parts could operate as models for humans. The question of the human emerges necessarily through his or her parts, and this is what I mean by the constitution of subjects *in* or via their parts.

Linda's assertion that many useful compounds correspond to solutions that could "not possibly work in humans" reflects the challenges that exist for an enterprise reliant on animal models that constitute a stand-in for the human. After one panelist's discussion about mouse models in research on schizophrenia, an audience member asked, "Can mice be schizophrenic?" The question compelled the audience to laugh, though it was unclear whether it was posed in jest.

However, the very possibility of *schizophrenia* research being based on the cognitive system of a mouse relies on the possibility of translation between the respective cognitive systems of humans and animals. Even when acknowledged by a scientist as merely a *model* or an *animal model*, an animal model's reference to the human is enabled through an epistemology of parts. The ability to establish a semantic tie between the experiment using a mouse model and the fictive schizophrenic human patient relies on the *isolation* of the mouse cognitive system as the means through which the human is tethered to the experiment and through which the human subject is *conjured up* at all. Here again, I suggest a constitution of patient subjects via their parts in these translational spaces.[4] Without such isolation, the remaining parts of the animal body become an analogical hindrance. By positing some sort of animal-human commonality (Friese 2013) between the cognitive system of an engineered lab mouse and that of the human—even under experimental pressures—we see parts as a necessary and entirely indispensable conduit for such a translation.

While Gail Davies in her work on the role of animal models in the larger translational imaginary adeptly discusses the humanization of the laboratory mouse in the form of the engineered laboratory mouse, the interchangeability that she notes reveals a reconstitution of the human as well. Thus, thinking with Davies, the translational capacity of the humanized mouse to stand in for the human in biological research models *can only work* in a context in which humans could be modeled or conceptualized in and via their parts—a parts-based epistemology. For example, the very operability of the notion of translatability between mouse immune systems and that of humans, or the very possibility of trying to reproduce cognitive stress using a humanized mouse in a laboratory setting, relies on a partial constitution under experimental pressures of the human subject. One must sideline all of the other aspects of the human body, its unique changes, as well as the complex experiences and the conditions of a disorder, to make the animal model make sense. At the same time, the very notions of therapeutic value and cursory references to patients and lives that one finds in translation-focused discussions indicate a referent to actual individuals. Thus, by the constitution of people via their parts, translational science and medicine help to produce a *partial subject*.

In much recent work on the transformation of the life sciences—the ways biological life has become an experimental domain—the biosciences have treated biological material (the human, the cell, the genome, or the molecule) as the primary unit of analysis. Thus, the panelists that I

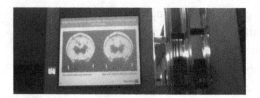

**Figure 5.10**
The 2011 Neurotechnology Investing and Partnering Conference at the St. Regis Hotel in San Francisco. The conference also included a visual panoply of fMRI images, microscopic images of cells, pathways, proteins, graphs representing clinical data, as well as pictures of brains.

witnessed discussed new approaches to psychiatric illness simply in terms of compounds, approaches, data, and models. Yet in their most detailed description of biologies, we also see something that has the elements of a subject. The metaphor of pixelation gets at the way that knowledge regimes such as molecularization may still actually help produce subjects. It is just that these subjects are only partial (see figure 5.10).

An ontology of partiality enables a reenvisioning of the demarcations of cells and systems, of redesigned genes and blood together as composing a sort of partially embodied form. It is the difference between viewing an image as a collection of dots or as unassembled dots, versus reading it as a partial rendition of a larger image. Thus, there is a difference between analytics that posit the complete erasure of a subject versus one that reveals the ghost of a subject: a subject left in parts, an assemblage of remnants. In this sense, a simple critique founded on biomedical, pharmacological, and technological reductionism or erasure may miss specific kinds of ontological constitutions that are at work in TN. Annemarie Mol's (2002) useful invocation of ontologies in her work on the way atherosclerosis is constituted multiply via spheres of practice offers a rethinking of the ways clinical encounters and their attendant worlds of practice inevitably produce multiple diseases. One could propose a similarly diverse set of practices that contribute to alternate versions of biological disorders.

What happens to the subjects who become attached to those disease ontologies? How does one act toward a partial subject? Of course, how one thinks of subjects has ethical implications: What social obligations do members of the industry or the public have toward subjects who are only partially constituted? Is this partial constitution responsible for the nearly sole focus on biomarkers and the neglect of the broader social, political, and economic conditions that produce the condition for which biomedicine is summoned as a totalizing solution?

Dealing with partial subjects or subjects in or via their parts also brings us back to the difficulty of the subject for TN. Indeed, the brain is a mess; unruly cells, turgid activity, and ever-changing complexity produce a storm of challenges for classic psychopharmacological models. Translating "the brain" also requires coming to terms with the unabashed messiness of people—of complicated and tethered patient subjects. TN is faced with the problem of brains that will not translate themselves: of coming to terms with complications of patient subjectivities, brains that sometimes don't respond to dosage models, and the inevitable individualization of human experience. In this way, TN is again about risk—scientific risk, patient risk, investor risk, and market risk. As seen both in its discourses and in spaces of scientific enactment such as the Neurotechnology Investing and Partnering Conference, TN also functions as a means for creating order out of biological, sociological, and technological chaos.

# 6  The Bedside: Patients and Pragmatics and the Promise of Health

During a quick exchange in 2017 about this project, in which I referenced the many financial architectures that inform TSM, notable STS scholar Sheila Jasanoff offered an important reminder: that TSM is "many things." Philosopher Miriam Solomon made a similar point during a quick conversation at a conference in California. The multivalency of TSM means that while it certainly involves clear political and economic functionality—and it is this functionality that I believe marks its uniqueness as a paradigm—it is the product of many histories and has many facets.

Yet in the translational paradigm, patients are always positioned as the ultimate benefactors of this massive intellectual and organizational restart. According to the NIH, the entire translational paradigm shift is about ushering in game-changing transformations in human health. In fact, much of TSM is justified in the name of placing patients at the front and center of scientific decision making about research and research directions. Scholars such as Alondra Nelson (2013) and Steve Epstein (2005) have shown the ways that community and patient activism have shaped both large-scale science and local healthcare innovations to address structural research inequities. Thus, there is a critical historiography of TSM that places patients and patient activism as a key driver of TSM (Arden et al., forthcoming). Indeed, the history of patient activism is part of the emergence of TSM. However, just as patients were "partially" present during the Neurotechnology Investing and Partnering Conferences, additional on-the-ground research shows some of the ways in which patient needs may not always be legible for the larger biomedical paradigms of TSM, and TN specifically. On-the-ground analyses reveal a reality above and beyond TSM's shiny discourses about patients.

While my larger research took me to investor partnering conferences, venture capital firms, universities, and neuroscience laboratories, a central space of investigation explored the worlds of the patient and of the clinic. One of my more instructive experiences in the field included time spent at the Cleveland Clinic in Ohio. I spent time with clinicians, attended clinician-focused meetings and events, and spent time with patients, hoping to learn about both their perspectives and their experiences with the conditions for which TN had been envisaged as a solution. The sections that follow detail these experiences and lessons learned about the marginalization of patient voices in TN as well as the complex relationship between TN and the clinic.

## 6.1   The Clinical Side of Translational Neuroscience

In a 2005 interview, then–NIH director Elias Zerhouni suggested that the move toward translational science and medicine was launched as a response to a single question: "What novel approaches can be developed that have the potential to be truly transforming for human health?" (Zerhouni 2005, 1621). Underlying Zerhouni's seemingly mundane question was something important, namely, the global ethical claims made around TSM—claims that percolate into powerful normative justifications for the massive investments, rearrangements, redirections, and institution building necessary under translation. At the same time that it is a locus of such claims, TSM ought to also be evaluated in light of such claims.

Though it is rarely touted, one important element of the translational imaginary of "novel approaches" is the "bedside-to-bench" translation. As one NIH departmental director, Francesco Marincola (2003, 1479), wrote, "Translational research should be regarded as a two-way road: Bench to Bedside and Bedside to Bench." In the founding editorial of the new *Journal of Translational Medicine*, Marincola (2003, 1479) wrote the following about the role of clinical medicine in the translational process:

> The purpose of translational research is to test, in humans, novel therapeutic strategies developed through experimentation. This concept is so popular that Bench to Bedside Awards were developed within the NIH to encourage collaboration between clinicians and basic scientists across institutes. But a more realistic approach would be to encourage opportunities to pursue Bedside to Bench research since our understanding of human disease is still limited and pre-clinical

models have shown a discouraging propensity to fail when applied to humans. Translational research should be regarded as a two-way road: Bench to Bedside and Bedside to Bench. ... Bedside to Bench efforts have regrettably been relegated to a Cinderella role because the scientific aspects are poorly understood by full time clinicians and the difficulty of dealing with humans poorly appreciated by basic scientists.

Marincola's acknowledgment about the actual difficulties of bedside-to-bench influence inspired several questions about the experience of clinicians working at the intersections of patient medicine and biomedical research. Anthropologist Michael M. J. Fischer chronicles the insights of Harvard medical scientist Judah Folkman, whose research on angiogenesis enabled the creation of several tumor therapies. In a piece on translational science, Fischer (2010, 339) recounts Folkman's sentiments about the ever-changing context that surrounds the physician-scientist subject:

Today the increasing difficulties of maintaining the dual career of physician-scientists, and the disjunctures caused by new discoveries ... [come] increasingly from basic scientists without much clinical experience and unaware of many important clinical clues. It is no accident that angiogenesis research [the discovery for which Folkman is most noted] began in a surgical laboratory (his), not in a molecular laboratory. In a molecular laboratory, like that of his good friend Robert Weinberg, he joked, one thinks of the life of cells that are flat, while surgeons like himself think that cells are crowded like they are in cancer tumors.

This excerpt brings us to consider the epistemological and pedagogical proposal that I make here about environments. Certainly, one's location in relation to patient subjects informs one's conceptualizations and practices. As Lesley Sharp chronicles in her ethnographic work about bioengineered prosthetics (2011) and devices (2013), the sheer distances between bioengineers and the worlds of patients create a critical and defining focus. Discussing the work of prosthetics bioengineers, Sharp (2011, 4) suggests that "unlike the transplant surgeons who will, in the end, employ their devices, bioengineers rarely enter cardiac wards peopled with bedridden patients." For her, this isolation translates into practice: "Social isolation from patients enables bioengineers to think abstractly about organ design and privilege 'hardware' over 'software'—that is, to foreground device concerns over implantees' lived experiences" (p. 4).

Sharp's analysis mirrors the ethnographic impulse about the pedagogies that uniquely emerge from ethnographic locations, as well as the influential cautions of feminist epistemologies such as standpoint epistemology

(Harding 1993) regarding the situated nature (Haraway 1990; Keller 1995; Longino 1990) of all knowledges. Not only are presumptions a part of the design of innovations, but absences and omissions are also similarly embedded. Because TN does not start from patient experiences, there is a similarity in the way that these *distances* inform the designing of TN as an intervention.

Because of this question about the relationship between an "innovation" and the people most affected by it, the need arises to understand the clinical register. I wanted to understand the ways that clinicians working within the sphere of brain-related illness make sense of the increased push toward the translation of neuroscience research into value for patients, to learn about the spaces where these therapies are rendered useful or usable. As clinicians are at the site of patients, this experiential knowledge was supposed to become a recursive flow of clinical data that would, according to the bedside-to-bench model, inform the development of therapies, thus turning the clinic into a laboratory.

Given their implication in this translational context, and the importance of social location and practices, I wanted to know about the experiences of neurologists and neurosurgeons. However, every neurosurgeon I approached flatly refused my request for an interview. Nevertheless, using several personal networks, I was able to interview two neurologists, one of whom I discuss below.

## 6.2   Conversations with a Neurologist

One prominent young neurologist, Dr. Jones, recently promoted to the head of neurology at a major hospital on the East Coast, offered to share her experiences about the clinical realities of trying to integrate a shifting environment with everyday clinical practice. For Dr. Jones, the desire to help people was a significant part of her decision to become a doctor. Working as a physician-scientist at a prominent university hospital, she was ambitious, sharp, and busy. I asked her a question about the translation of neuroscience from scientific laboratory to clinical spaces:

> There are two problems: one is with animal research and the brain—thinking about the brain as a complex organism. So therapeutics that are targeted at those models may not necessarily work the same in a human brain and sometimes it doesn't work in the mouse brain. The other is translating an effect in the

laboratory to an effect clinically. I can think of drugs that have been proven clinically to have a major effect on the way that dementia progresses but in reality it's not very significant. (Interview 8, May 29, 2010)

This problem of animal-human translation was a narrative that several stakeholders repeated—from physicians to bench scientists as well as executives at biotechnology firms. Yet the physician-researcher subject is also a role that is fraught and complex. Jones explained the complexity of working as both a physician and researcher and how the translational imperative influences her research work:

> I think that maybe in the past we were driven by our own curiosity and you didn't have to come up with products for a paper or have a goal in mind for research and now a lot of it is financial; you have to be thinking about how am I going to translate this for something that I can do for patients—and that's a difference. Because of the lack of funding and resources, people are having to come up with something for their research. [Before] there was more intellectual freedom but discovery wasn't necessarily the goal. (Interview 8, May 29, 2010)

I asked her whether her research had impacted her own practice thus far. Had she found a relationship between her own personal translation-focused research activities and the way she practices neurology? She said that she had not felt an impact. I was thinking about the bedside-to-bench translation that was to be a solution to the problem of the clinician-scientist gap when she modified her answer. She explained that scientific discourses did in fact impact the clinical experiences of patients. For Jones, science was useful as a means of garnering trust via the authentications of expertise. She described the way that neuroscience has become a source of clinical explanation and a means through which to compel adherence on the part of patients:

> I think without knowing the why of a disease, you can't ... it's very hard to have a viable field and viable practice and patients' trust, without really knowing the why of a disease. In talking to patients they are a lot more impressed and more likely to follow recommendations if they [the physicians] say, "This is what you have" and "This is what's causing it." In the past we didn't know why and so now you are able to deconstruct a lot more. It's really, really critical for the field of neurology to continue to advance and without neuroscience it wouldn't be able to advance in the way that it has. (Interview 8, May 29, 2010)

In thinking about this pre-neuroscience version of neurology, I was interested in her use of the term *deconstruct*. I took this to mean a discursive deconstruction—an ability to provide a better narrative about patient's condition, which, for clinicians and patients alike, is comforting. Scholars have pointed out the importance of biomedical knowledge for providing comfort (especially given the discomfort of not knowing). The soteriological element at work in Western biomedicine means that both patients and practitioners cling onto hope through biomedicine's explanations and solutions (Good 1994, 86) in ways that tie patients to both desires for understanding and faith in the larger biomedical enterprise.

Yet, as Jones continued to expound, she began to articulate the ways that the values she had just articulated—values predominantly about information and explanation—are not the same as improved therapeutics and better outcomes. The complicated role of information in the biomedical encounter is made clear in the complicated morass of diagnostics. As we spoke, she continued to think through and deconstruct the impacts of an increasing emphasis on diagnostics for her neurology:

> The cost of certain technologies is a problem, and even though now we're able to make a different diagnosis, it doesn't change what we will do; we spend a ridiculous amount of money diagnosing something and the outcome for the patient isn't any different and that is the biggest issue. There's more emphasis on diagnosis and not more [emphasis] on treatment. We're able to have an exact diagnosis, but a lot of healthcare dollars are going where it isn't changing the outcomes. (Interview 8, May 29, 2010)

Here we are made to think about the effects of the corporatization of healthcare as it meets the utilization of the life sciences in clinical practices. After thinking a bit, Jones then delved into greater detail about the kinds of diagnostics that are part of her everyday clinical practice: "We do different kinds of neuroimaging as well as different kinds of antibodies for different diseases and we didn't have the antibody test before. These are the two biggest, and we spend a lot of money on the diagnostic MRIs, fMRI, and we send blood tests to all kinds of places because those things are now available."

Taking her statement about availability literally, I realized that diagnostic testing increases simply because of its mundane availability. I also came to realize how a technical orientation toward therapeutics increases the use of technologies, even when the outcomes may not be necessarily and demonstrably improved—an issue that several scholars working on neurotechnologies have highlighted (Joyce 2008; Kesselheim, Mello, and

Studdert 2011; Leibing 2009). Increased diagnostic testing and diagnoses bereft of tangible benefits in terms of patient outcomes may connect with what Joseph Dumit (2012a, 155) calls *surplus health*, which he defines as a means by which biopharmaceutical companies profit from prescription maximization without benefits necessarily accruing to patients.

Indeed, this trend reflects the accumulative nature of pharmaceuticalization, and thus it is unsurprising that massive new knowledge infrastructures would coincide, as Jones reported, with little shift in terms of outcomes. Dumit suggests that science infrastructures are turning toward the production of research that supports and pushes *increases*: increased prescriptions, increased diagnostics, promises of lifelong pharmaceutical entanglement. Yet Jones, who astutely explained the nuances and pressures that affect modern-day clinicians, is also aware of the trend toward consumer empowerment through personal research and self-advocacy, which is an important part of the picture. But how do we account for the bioscience work that physician-researchers undertake to create new knowledge about epilepsy? For Jones, who often expresses her commitment to her patients and her research, information is in itself a good thing. She explains the way that the diagnostic test is easily available, though not always easily usable:

> I've heard discussions … [from] neuromuscular neurologists—with their frustration with the different laboratories and how easy it is to order a test when it's difficult to interpret; when things are very specialized, you need to know how to interpret the test. Some things are easy—if they are positive or negative—and then some things are more complex. (Interview 8, May 29, 2010)

It was from this point on that I began to think about TN in terms of its capacity to *attach disorders to biological material*. The way that the diagnostic test indexes the meeting of the scientific laboratory (or laboratories) and the clinical encounter is a relationship that is difficult to map. To the extent that neuroscience enables greater biological meaning making around brain disorders, translational medicine may enable the production of "patients-in-waiting" (Dumit and Greenslit 2005), but it's also about a kind of emergent materiality: the production of material-discursive worlds that can be leveraged as a technology of trust for desperate patients and also monetized by Big Pharma.

## 6.3   Realities for Patients and Health

I met Shana during a patient education session at the Cleveland Clinic, an eminent academic hospital and research center in Cleveland, Ohio. Shana came to the session to learn more about epilepsy since her son has a particularly difficult strain. She attended a support group for families of children with a clinically challenging form of epilepsy, also at the Cleveland Clinic. Epilepsy is a brain condition marked by recurring seizures (convulsions). In the United States, 467,711 children have epilepsy,[1] the severity of which can range from mild to extreme. For those with extreme symptoms who are pharmacoresistant, clinicians may recommend surgery and other invasive therapies (Theodore and Fisher 2004).

"Jawan had been on medication for years," Shana told me. "But the drugs just didn't work. We went on this medication and that one. I was so scared, and so then they transferred us here." Once she confided to me that she was afraid to keep hoping—a cumulative exhaustion from the constant worrying and searching for answers. Shana is an African American woman in her forties. She used to work as a home health aide until around 2006, when she lost her job at a local Ohio health worker agency.

Perpetually shy and quiet, with a young face, Jawan appeared to be much younger than his actual age. My questions to him would usually go unanswered beyond a mere shrug, rarely compelling him to emerge from hiding behind his mother. Despite his age, he never uttered more than a few words to me in all of our time together. Jawan had his hands wrapped around her on every occasion that I saw them. He would clutch her, holding his hands around her waist or around her arm.

The Cleveland Clinic has been a primary source of jobs and healthcare in the city of Cleveland—an especially salient fact given that the institution is in a blighted and transitioning neighborhood. In fact, it sits directly on a sociopolitical boundary line. On one side is middle-class housing inhabited primarily by professionals who often work at the clinic, and on the other side are poor, blighted neighborhoods. One experiences the difference just walking from one part of the center's building clusters to the other.

The Cleveland Clinic is not only one of the largest employers in Cleveland, but it is also a beacon of excellence in terms of healthcare, research, and education. Its federally funded research and nationally renowned clinical programs attract money into the region. Often compared to the Mayo

Clinic in Minnesota, the Cleveland Clinic occupies an important political and economic role in a city that has been ravaged by a lack of high-paying jobs and by political turmoil.[2] According to the 2010 American Community Survey,[3] Cleveland was the second poorest city in the United States behind Detroit.

Shana had worked on and off for all 13 years of her son's life. However, the severity of his epilepsy meant that she had to be, in her words, "always on and always ready." The toll that epilepsy took on her and her family turned her into a constant advocate and a perennial participant in the research trials in neurology at the Cleveland Clinic. Jawan had not experienced a seizure in several months, but they were unpredictable. Our conversation, which we sometimes had at a nearby McDonald's, often centered on how desperate she was to find a solution.

In 2010, Shana and Jawan were referred to the Cleveland Clinic's well-known neurology department and active epilepsy research arm. The clinicians provided an option beyond the typical solution of antiepileptic drugs. Typically, individuals not responsive to antiepileptic drugs are considered for brain surgery. At first, Shana was excited by the prospect. She began to research brain surgery: "I really thought that this was possibly going to fix it and that this was going to be different. I mean it's his *brain*" (her emphasis).

Neurosurgery for epilepsy works on the premise that performing surgery on the part of the brain producing seizures can reduce or even eliminate them. However, even surgery is not a guarantee. Some patients do not improve with surgery, and since they are often pharmacoresistant, they are at a loss in terms of options. It was at the Cleveland Clinic that Shana learned about Deep Brain Stimulation (DBS), which research had suggested might be an option for patients whose condition was pharmacoresistant.[4] As previously noted, DBS is a special kind of brain technology in which a pacemaker-like device delivers electrical stimulation through electrodes implanted in the brain. The installation of the device is dramatic. Surgeons create small holes in the skull to implant the electrodes and perform chest surgery to install the battery. At the time, DBS was used primarily for Parkinson's disease. However, increasingly and controversially, it's being pushed for use in psychiatric cases, including depression and obsessive-compulsive disorders such as obesity (Fins et al. 2011; Throsby 2009). Medtronic, the device's manufacturer, has worked aggressively to expand potential markets for DBS.

DBS is often positioned as a model case for TN (Kringelbach et al. 2007), a model example of how TN might be able to produce positive health gains. And yet the technology embodies significant risks as well. DBS represents a clear two-way relationship between observations about brain functioning and the possibility of turning that knowledge into immediate therapeutic impact. It is both therapy and (when used along with magnetic resonance imaging) a means of research. In short, it is an avenue by which one can both manipulate neural networks in the brain in novel ways (Kringelbach et al. 2007) and, ostensibly, view the fundamental mechanisms that underlie human brain function. In this regard, DBS can be used to learn about the relationship between disease and brain states.[5]

Jawan's physicians presumed that he would be a good candidate for DBS, but Shana, who had researched the possibilities independently, had concerns. She said, "Even the testing required to know if Jawan would be a good candidate [for DBS] sounded scary. They called it invasive testing, and they'd open his brain up. The doctor says that there's research and that the testing is worth it if we ended up using DBS."

Then Shana leaned in to share something with me. During this discussion, she and I were sitting in the foyer of one the newest buildings on the Cleveland Clinic's campus: a minimalist seating area with white floors, white walls adorned with light-colored art, and modern indirect fluorescent lighting. The result was a space that was both calming and sterile, somewhere between an art museum gallery and a laboratory. At every corner, there were hallways that led to various parts of the hospital; however, the space itself was acoustically muted.

When close to me, Shana revealed that she didn't feel the clinicians were certain about whether DBS would fix things. She said the neurologist admitted that although DBS could be beneficial, the testing, which requires invasive probing using depth electrodes in the brain, creates its own risks—a process too risky if not offset by the benefits of DBS.

This added layer of risk was acknowledged openly during a symposium at the Cleveland Clinic that I attended on October 3, 2010. During one session, Imad Najm, leader of the Cleveland Clinic's Epilepsy Clinic, said, "I try to balance the benefits with the negative" when thinking about the consequences of invasive brain testing. During his presentation to the audience, he posed the following question: "Are these things worth the risks that we're taking?" Of course, the question of risks produces new questions:

about risks for whom, about the temporal contingency of risks, about risks and patient experience.

Several presentations at the symposium—attended by neurologists, neurosurgeons, and nurses, underscored an unspoken lack of information on potential risks for patients and for physicians. The risks include electrical toxicity from the electrodes and lesions on the brain at the site of implantation. As a new therapeutic option, DBS for epilepsy represented a clear example of how increased knowledge of the brain, especially in terms of diagnosis, translated into potential health gains but also into significant risk (Kringelbach et al. 2007).

While considerations of how much to share with patients were mentioned, the symposium focused largely on how much was simply *not known*—about DBS, about invasive testing, and about the brain itself. Mark Bernstein, a neurosurgeon and oncologist, articulated the dilemma this way: "We do so many things and make so many decisions without class 1 evidence and in neurosurgery, it's [something like] under 10 percent of decisions [that are] made with class 1 evidence for stuff that we do and it's an ethical blight on neurosurgery." This ethical blight includes the "deception" that occurs in hiding the information ambiguity undergirding clinical decision making from patients and from the public. Beyond sheer deception, the ethical blight encompasses countless risks neurosurgeons take because of a dearth of empirical knowledge. During the same symposium, Steven Schiff, a neurosurgeon and professor at the Cleveland Clinic, stated, "We don't think about the biology of placing soft material onto the brain." Schiff ended with the suggestion that risk-benefit analyses should "incorporate greater knowledge of biology."

The risks of invasive testing for DBS and the many questions that exist around this technology emerged as a primary theme. Toward the end of the first session, one audience member posed this question to everyone in the room: "What happened to the most severe patients who chose not to do the surgery or who were not approved?" A second person posed another question: "How do we know what works in the absence of data from alternative therapies?" How does a lack of biological knowledge create specific new risks? Given the vast unknowns about what DBS actually does in the brain or about how the brain truly and precisely functions under stimulation, DBS is *an uninformed performance*. In this space the feedback between knowledge and biotechnologies comes to the fore. Concerns abound regarding the knowledge that created this technology but also how this technology creates knowledge.

DBS, first marketed in 1997, is an early example of a "successful" neurotechnology and it was produced and commercialized by the large biotechnology firm Medtronic. Medtronic is a major player in biotechnology device research and design and a major player in the area of drug-delivery devices. After DBS was approved in 2009 for the treatment of Parkinson's, Medtronic pushed it toward additional markets, including epilepsy. Problems regarding DBS for epilepsy appeared early: in 2010, the Food and Drug Administration (FDA) raised concerns about the safety of the DBS device and a clinical study failed to prove that patients using the device experienced fewer seizures than those in the control group (Lee 2010). The FDA voted against approval in 2010 as a treatment for epilepsy. Its use in the United States became investigational, while it was approved for wider use in Europe, Australia, and Canada (Walsh 2013a). In 2018 Medtronic won FDA approval that enabled DBS to be used for epilepsy patients in the United States. However, in the years prior, concerns about the risk of DBS permeated "on-the-ground" discussions about the technology.

## 6.4   Risk and Exemption

In May 2013, the FDA cited a Medtronic device (Cortez 2013) for flaws related to its installation and issued its highest-priority-class recall.[6] My interlocutors—both biotechnology investors and people recently employed at Medtronic's Silicon Valley location—didn't seem fazed during phone calls about it. John Chen, someone with whom I developed a relationship during years of attending the industry conferences, told me that there had been rumblings for a while about "the wires in those devices." According to Medtronic, the problem was resolvable through a change in manufacturing process and physician-directed installation instructions.[7]

So, if DBS failed to get approval in the United States in 2010, how could it have been offered as an option to Shana and Jawan later that same year? The reason is that Medtronic applied for and received appproval to use DBS for non-standard uses under a set of exceptions. The first exception is a designation called "investigational use," whereby companies can use an as yet "unproven" technology for "investigational purposes." DBS for pharmacoresistant epilepsy was considered investigational. The second exception is a strangely named exemption from the FDA: the *Humanitarian Device Exemption* (HDE). Under the HDE, companies like Medtronic bypass the normal FDA requirements for clinical trials using standard measures and size in order to benefit smaller patient populations. The HDE program is the device-focused counterpart to the Orphan Drug Designation program. Both the Orphan Drug Designation program and the HDE program are administered under the FDA Office of Orphan Products Development (OOPD). Since its inception, the FDA's Humanitarian Use Device Program has enabled first-step approval for more than 50 technologies (Food and Drug Administration n.d.).

To understand the exemptions and context of HDE, I provide an excerpt from the FDA program description that outlines the mission of the OOPD:

> The FDA Office of Orphan Products Development (OOPD) mission is to advance the evaluation and development of products (drugs, biologics, devices, or medical foods) that demonstrate promise for the diagnosis and/or treatment of rare diseases or conditions. In fulfilling that task, OOPD evaluates scientific and clinical data submissions from sponsors to identify and designate products as promising for rare diseases and to further advance scientific development of such promising medical products. The office also works on rare disease issues with the medical and research communities, professional organizations, academia, governmental agencies, industry, and rare disease patient groups. OOPD provides incentives for sponsors to develop products for rare diseases. The program has successfully enabled the development and marketing of more than 400 drugs and biologic products for rare diseases since 1983. In contrast, fewer than 10 such products supported by industry came to market between 1973 and 1983. The [more targeted] Orphan Grants Program [within OOPD] has been used to bring more than 45 products to marketing approval.[8]

The primary goal of OOPD is to advance development of products through the creation of incentives. In the description below, taken from a symposium report on HDE (Kaplan et al. 2005), note the efficacy-related exemptions and importantly the role of HDE (sometimes referred to as Humanitarian Use Devices or HUD) in the creation of a pathway toward an otherwise complicated market access:

> The HUD/HDE pathway was created by Congress to facilitate the availability of medical devices for "orphan" indications, i.e., those affecting <4000 individuals within the United States each year. The HUD/HDE pathway streamlines the approval process and permits less well-characterized devices to enter the market. HDE approval focuses primarily on issues of safety and scientific soundness and

does not require demonstration of efficacy. In the 7 years since the first device was approved in 1997, a total of 35 HDEs have been granted (23 devices, 6 diagnostic tests). As the costs to gain regulatory approval for commonly used devices increase, companies often seek alternative ways to gain market access, including the HUD/HDE pathway. (Kaplan et al. 2005, 502)

The FDA created the HDE to incentivize companies to bring biotechnologies to market for populations too small to make such investments financially worthwhile for pharmaceutical and biotechnology companies. This mode of *sidestepping* or *fast tracking* constitutes a utilitarian version of ethical variability (Petryna 2005), which exploits the "problem" or crisis of a small market size as a justification to allow device developers and manufacturers to bypass risks of clinical trials. This illustrates the framing of medical issues as market problems that inhibit biotechnological innovation. It is only within such a frame that recognition of small market populations makes sense.

The very naming of this regulatory pathway—The Humanitarian Device Exemption—invites questions about whether this pathway reflects what French anthropologist Didier Fassin (2012) calls "humanitarian reason." He uses this concept to make sense of political substitutions achieved through the mechanics of compassion created by and authored in the name of humanitarianism. In his case study, Fassin discusses the special status accorded to children with AIDS in South Africa. Similarly exceptional subjects are reified and invoked in the case of HDE as individuals suffering from "rare diseases." In both cases compassion can be operationalized within the given cultural context to enable access to special kinds of exemptions. Following Fassin, I ask whether these appeals to compassion supersede appeals to morality, justice, and other rights.

In an analysis of the moral machinery of HDE, the notion of "humanitarian reason" is most useful because of the way it traces the mythopoeia that surrounds intervention. Humanitarianism enables important social and political obfuscations: Fassin's case study shows that the sentimentalization of children with AIDS caused a reification of those subjects in ways that leave unaddressed the sociopolitical contexts that give rise to such realities. Thus, humanitarianism allows one to "aestheticize "compassion "at the same time as implementing policies that increase social inequality" (Fassin 2012, 2).

The relevance of this element of humanitarianism's functionality for HDE is multifold. As scholars have chronicled (Biehl 2007; Benjamin 2014; S.Epstein 2005; Martin 2007; Wailoo 2017), NGOs, patient-activists, communities, and detractors have fought for resources that are reflected in present-day interventions and programming such as those created at the NIH and FDA. The activities of such groups were especially crucial in achieving federal recognition for AIDS in the United States and in securing AIDS research funding.

Yet, that HDE is positioned as a solution to the problem of orphan diseases conceals broader issues related to the market.[9] As HDE is an intervention built on the need to create incentives for private companies to invest in the development of biotechnologies for smaller market segments, it reveals the way that market logics become *the means by which* biomedical interventions are justified, structured, and organized. The very nature of healthcare corporatization produces the conditions that render this form of regulatory exception to be enunciable in the language of humanitarianism.

The case of orphan drugs illustrates how moral sentiments emerge from the needs of market logics (Biehl and Petryna 2011) to create states of exception, bypassing patient protections in order to incentivize risk-laden research. The issue of transferring risks onto patients and even, in the context of the FDA, onto the "public" in the name of health brings up questions about the public-private distinction in governance/regulation/protection.

Shana ultimately declined the DBS option for Jawan. She offered many reasons; however, principal among these was what was then a lack of long-term research about the safety of DBS for epilepsy.

In their work *Risk and Culture* (1983), Mary Douglas and Aaron Wildavsky highlight how risk is not merely the natural response of individuals and collectives to a lack (or overabundance) of information—and that there is more to the process of risk selection and perception than the presence or absence of information regarding harm. Importantly, they outline the way risk perception invariably involves social processes and values, thus denaturalizing risk as somehow inevitable, uncontrollable, and natural. Existing risks often entail a complex of prior decisions and preparations or the lack thereof. A constructivist view of risk enables one to see how risks emerge from decisions about knowledge and nonknowledge.

DBS was approved for use in patients with refractory or drug-resistant forms of epilepsy in 2018 after additional long-term research about its use in patients with epilepsy. Yet, this approval did not resolve all concerns about its use. Given that DBS is touted as a model form of TN (Kringelbach et al. 2007), and it stands as one of the few "products" to emerge out of the translational shift, many will look at DBS in assessing the success of TN in the long term.

Yet, the larger question about the acceleration of R&D under translation and the implications in terms of "too-early" development remain. Is TN especially conducive to epistemic risk? Further complicating the issue of risk is the indeterminacy introduced by the inherently communicative nature of translational science (Davies 2012)—composed of scientific, social, technological, and material processes in interactions that create a field of potential fissures and mistranslations. In his essay "Lively Biotech and Translational Research," anthropologist Michael M. J. Fischer (2012, 388) identifies the many metatranslations embedded "across science fields and technological scales: from bench to clinic, ... from green fingers to stable techniques and scalable production; and from experimental therapy to standards of care." At every point of biological, material, and social translation, the opportunities for risks are born, inevitably part of the emerging worlds (Fischer 2003) at work in contemporary knowledge societies.

One useful example of the way that translational science is a particularly conducive space for the production of epistemic risks is pharmacology's reliance on the research mouse. Similar to *Drosophila*, the research mouse is an essential experimental stand-in for the human. Of course, this cross-species translation is highly fraught and a known problem. In her research on the epistemic and spatial roles of the humanized mouse in translational research, Gail Davies (2012, 132–133) recounts a useful contemporary example of the risks inherent in presumptions of corporeal and environmental equivalence:

> Periodically, an event dramatically reveals the gap between experimental model and human corporeality. The most recent was the catastrophic failure of preclinical safety trials at Northwick Park, in north London, in 2006. ... The trial was testing a drug, itself a humanized monoclonal antibody, being developed to treat leukaemia and rheumatoid arthritis. The drug had been tested on mice, with no adverse consequences. Yet the first introduction of a small dose of this compound into humans produced sudden and systemic organ failure in all six trial

participants not taking the placebo. Monoclonal antibodies are biotherapeutics that bind to specific cells in the patient's body. They are not the therapeutic agent themselves, but stimulate the patient's own immune system to attack the targeted cells. In this trial, the participants' immune systems overreacted, attacking the body indiscriminately in a so-called cytokine storm. The drug was removed from development, the failure of the trials investigated, and the six men warned to expect a lifetime of health problems.

The important point here is that TN's impact on laboratory practices can also exemplify the presumptions of equivalence that can constitute good examples of a breakdown in the translational continuum. Yet TN can also exemplify how knowledge enfolds epistemic risks and how interventions such as DBS for epilepsy rationalized on the basis of incomplete knowledge contain these risks:[11]

> The mice used for safety testing, kept in sterile laboratories, had not had the same immune challenges as the human trial subjects. Their inexperienced immune systems did not react. In humans, the memory of past infections contributed to the detrimental response. The species, but also the spaces, were not representative. (Davies 2012, 133)

A social constructivist/denaturalized view of risk provides a useful way to view not only increased risks from potentially premature biotechnological developments, but also risks that emerge from the kinds of exceptions provided to experimental biotechnologies via Humanitarian Device Exemptions. It suggests that creating biotechnologies based on problematic knowledge *is* already a social decision about risk. Here we can think of this social tableau as being composed of bioengineers, investors, patients, clinicians, and regulators—all implicated in an ethos in which biotechnologies get understood as the most important means of intervening in health.

The example of HDE shows how innovation in biotechnology often relies in practice on systems of *transferring* risk. In the case of DBS for epilepsy, HDE enables risks to be passed onto patients via the exemption for a clinical trial; risk is transferred from the biotechnology company onto the patient and therefore the public. While the ethics of HDE are attracting scholarly attention (Alpert 2011; Fins et al. 2011; Peña et al. 2007), there is little conversation about how TN engenders all kinds of new risks—including those for patients. Here questions emerge about how TN might uniquely structure and restructure risk as part of the rush toward commercialization and how the ethos of translation may resignify these risks. Now that biotechnology companies are able to use HDE exemptions as part of a business strategy (Fins et al. 2011), the transfer of risk (to patients in the case of therapies based on incomplete clinical or therapeutic understanding) can be the bedrock on which a new mode of biotechnological innovation, a global turn to pharmaceuticalization, and increasing corporatization of health rest.

## 6.5    Risk and Translational Thinking

There is something instructive in the "technological drama" (Pfaffenberger 1992) in terms of *thinking* about risk and biomedicine. Shana's own thinking—which was not immediately about technological and medical possibility or the biomedical promises that haunt preoccupations of biomedical futurity, but rather about Jawan and his safety—compelled me to consider several points.

The first is about the conceptual worlds enfolded within the practices of TN. Is risk essential to the experimental system on which commercial neuroscience innovations rely? Rather than a mere product or byproduct (in the optimistic view), is there a more deliberate configuration not necessarily willful, but *inherent* in the interstitial actions, inactions, and microactions that are part of translational imagining?

A second issue is about how TN operates as a "vision" that does not begin with the experiences of patients. It brings into view the thinking that informs TN and that impacts its output. Figure 6.1 represents what could be considered a conceptual archive created by UCSF's Clinical and Translational Science Institute. The graphic instantiates one institutional vision regarding the translational continuum and its constituent parts. It shares with other institutional visions its *molecular* starting point. Therapeutic targets refer to the biological targets that are or can be manipulated toward a therapeutic effect. Even in this model, one sees the articulation of therapeutic or pharmacological thinking. Pharmacological or device-based endpoints are the pretext from which research designs are initiated. The productivity of investments in research programs gets measured in terms of the "efficacy" embedded in translational progressions such as this one. Legal preparations for commercial collaboration are framed in these terms and experimental questions are developed within the same frameworks. The challenge here is that these endpoints are not self-evidently the same as

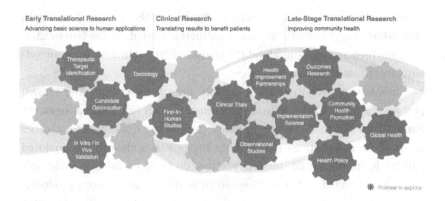

**Figure 6.1**
UCSF's Clinical and Translational Science Institute's interactive graphic. This image, a screenshot from the institute's website, outlines the relationship between all of the disciplines and stakeholders that are part of the translational research continuum. This also represents an elucidation of the various stages of research, product development, and applications. See http://ctsi.ucsf.edu/our-work.

better health—that is, unless that which is currently understood as health is evolving.

Nevertheless, this pharmacological or therapeutic pretext not only constructs the very objects of investigation, it also constructs its subjects. In figure 6.1 one sees a literal configuring of inputs and outputs. Although they represent key stakeholders, one notices a lack of community and patient input at the early stages within this model of the translational continuum. Translational research as depicted here begins with *hypothetical people* or, as I suggested earlier, partial subjects, whether via transgenic imaginations at work in the use of humanized mice models or in the elision of patients/ people at the outset of the translational continuum.

In some ways, translational research works through a set of critical constrictions—a winnowing down of research questions and agendas (Jogalekar 2012); a closing of non-application-focused federal funding (Reed et al. 2012); a narrowing emphasis on the therapeutic wing rather than the entire R&D process, including basic science (Minogue and Wolinsky 2010) and implementation science (Solomon 2010); a focus on fewer questions; and importantly, an acceleration of the discovery processes toward clinical trials (Garcia-Rill 2012). Such constrictions and their subsequent knowledge pathways are part of the sorts of mistranslations that occurred in Davies's example about the failure of preclinical safety trials in north London in 2006—knowledge constrictions or constructions[12] that, in the latter example, informed assumptions about the translatability of human immune systems and those of humanized mice.

I invoke this not in order to simply or naively critique the known problems at work in the reliance on animal models in biological research. Rather, I invoke this example in order to pose the question: What other kinds of stand-ins are at work? What logics and omissions are enclosed within the translational imaginary? How do such knowledge enclosures become embedded and mythologized in translational science's products? In a context in which molecular "therapeutic targets" are its primordial starting place, translational science can be understood as producing moral worlds—of placement and displacement, reification and sidelining, temporal expansions and compressions.

## 6.6   Pauline and Terry

Even the vague target of "human health" looks different from the patient's perspective. While clinicians and researchers train much of their efforts on diagnostics, modestly effective pharmaceuticals, and risky device therapies, patients in the throes of debilitating conditions might choose, and in fact benefit more from, sidelined solutions. As mentioned earlier, one such solution was touted by Dr. Stein, the Santa Cruz psychiatrist, who proudly noted his own stubborn refusal to abandon counseling in favor of drugs. According to Dr. Stein, "In ... working with patients, you quickly learn that the issues that people present can't be neatly fixed with some magic bullet." Stein's insight is that the constraints born of translational science and medicine aren't always in patients' interest.

Pauline was difficult to talk to, Terry less so. Our socially conscripted roles permeated our exchanges. We may have been an odd pairing in this particular part of Cleveland: I, a young, tall African American male anthropologist from somewhere far away, and Pauline and Terry, third-generation Clevelanders who, at first glance, embodied the white working-class section of Cleveland where they had lived their entire lives. Pauline said that she had been looking for work since 2001; "there aren't a lot of good jobs here nowadays," she told me. Terry's disability benefits meant that he received a check every month from the US government. I explained to them who

I was and the nature of my research. In our first meeting, Terry asked me where I was attending school.

"Princeton. Is that in Boston?" Terry asked.
"No, it's in New Jersey. You might be thinking of Harvard," I responded.
"Oh! He said, nodding politely."

The McDonald's restaurant where we often met was chaotic and convenient. Next door to the Cleveland Clinic, it was so frequently full of patients, clinicians, and staff that it was an extension of the Cleveland Clinic facilities. But my discussions with interlocutors took on a different feel and timbre at the McDonald's from the discussions inside the clinic. Terry, Pauline, and I would meet there for lunch periodically after a special patient support group that Pauline attended had concluded. Pauline often played with her cellphone during our lunches. Terry would always get the same thing: a Quarter Pounder Value Meal without onions and a large Sprite. Despite our initial meeting, I was always "that Harvard guy" who had lunch with them. I did not try to disabuse them.

For his 44 years, Terry was a well-built and strong man. He was 6 foot 6 and over 350 pounds. He had a tussle of strawberry-blond hair. His partner of 25 years, Pauline, was his primary caregiver. "My mom and sisters are all over the country," he told me one day. Terry often wore a baseball cap sporting the name and logo of the Cleveland Indians, the local sports team. He was boisterous and dynamic, and his hat sometimes became dislodged during moments of jubilance. His family was strewn across various parts of the country and without the financial resources to lend a helping hand, according to Pauline.

While Terry tolerated my strange ethnographic interest in his visits to the Cleveland Clinic's Epilepsy Center, Pauline was understandably skeptical and cold. In the larger context of their search for answers and solutions for Terry's epilepsy, my mundane questions felt small and perhaps inopportune—at least in the context of an everyday life that was rendered apocalyptic by Terry's violent seizures. The seizures dominated daily life for Terry and Pauline until, after one particularly horrific attack, they found an unconventional solution.

"So, you know those commercials with that old lady who says, 'I've fallen ... And I can't get up?!'" Terry's imitation of the actor in the commercial was perfect. For a moment Pauline smiled, on the verge of a giggle. I did remember it. That commercial had become a source of humor for its overly dramatic acting and the singsong quality of the commercial's main tagline. I recall many instances during middle school when classmates and I would imitate that dramatic tagline during recess. Terry explained its relevance:

> Well, that thing is called a Life Alert. It seems like it's only for old people, for the elderly and such, but it works for epileptics, too! We ordered one after I had a seizure back in 2008 in April. I had been at home because I can't work and I was taking a shower and all of the sudden, I was having a seizure. I don't remember everything, but I had been in the bathroom, and I had apparently fallen down and cracked my teeth on the side of the bathtub. Pauly came in and she had to call EMS. I had knocked everything down; the towels, the shower curtain. I was bleeding. I totally could've died. So, then we got Life Alert and so I have been able to use that, especially when I'm alone. (Interview 18, September 2, 2010)

Life Alert® is a personally wearable device that enables the wearer to press a button to alert paramedics of the need for rescue or emergency attention. Used widely by older people in the United States, its operative benefit is, perhaps ironically, the way it stands in for momentary lapses in social support. When Pauline was out running errands or on job interviews, Life Alert provided a means of immediate and proximal care.

"I couldn't do anything. I mean I couldn't do *shit* [her emphasis]. I mean I loved him—of course I love him, but it was so hard." She paused, searching for words. Her body communicated years of angst, guilt, commitment, and confinement. "It impacted everything we did. Like, I needed something that could give us some sort of *life*. I just couldn't be there every second of every day. You know?" (Interview 18, September 2, 2010).

Of the primary concerns for Pauline and Terry that emerged during our interactions, few were focused explicitly on emerging brain technologies. I also did not find them to be mired in an endless web of neuroscientific facts. In fact, knowledge itself constituted a hindrance—or more explicitly, merely a secondary source of exhaustion. Pauline's affect and mode of being spoke of perpetual exhaustion; it was difficult to discern her own disposition and personality from the years of weathering the epilepsy and its emergency modes. While my experiences with Jawan and Shana revealed the imprint of Shana's technological and clinical literacies, Terry and Pauline seemed permanently fixated on everyday existence. In Pauline's descriptions about post–Life Alert life, one sees that it constituted a technology of and for everyday survival. Few of my other ethnographic questions mattered to Pauline; she seemed disinterested in discussions about long-range or short-range futures.

One day, Terry and I talked alone about the toll that epilepsy had conferred and his sadness not simply at his own condition, but for the way it affected Pauline. "She's really a beautiful person," he said, perhaps sensing, rightly, that I couldn't know her well enough to have any sense of who she was apart from our interactions. "She's really been there for me and this whole situation [of epilepsy] just made her into a zombie. She couldn't sleep. She would cry and cry every time that I had an attack. I mean, you know, it would really scare her. And I really had no one left. It's not like I had this huge family. I was by myself. There were good times, too," he said, as his tone changed slightly. He continued:

> I really don't understand epilepsy all that well or any of the medical stuff either. I was just like okay, "Here are some pills" ... okay, "Now, here are some new pills," and Pauly, she's really good with that stuff. She'd keep all the pills together for me and stuff. I don't know what I would have done without her. She's definitely my saving grace ... like, a guardian angel. But she can also be kinda ... kinda bitter and grumpy. But it got better after I got this Life Alert thing. (Interview 18, September 2, 2010)

Terry caressed the imprint of his device, which was attached to a necklace that he wears under his shirt. "She can go and do stuff now and I say to her, 'go out and have a good time,' you know. Like, go get some cigarettes or something simple like that. It's been such a good thing. But we're still scared. I still gotta' be careful." Terry said that he had only had a single seizure that year so far and about three seizures the year prior.

My experiences with Terry and Pauline were instructive in unsettling several types of deterministic presumptions. I was struck by the way that their commitment to living a good "everyday life" related to the ways that epilepsy had impacted their "everyday." For them, Life Alert's utility was in its restoration of time, its marginalization of a kind of apocalyptic contingency that dominates life when living in a constant state of potential emergency. Their situations became the data for their own sense of immediate therapeutic value.

The reams of biomedical data or the clinical confusions did not arrest them as I had expected. Their lives had been much more haphazard and importantly, much slower than could be predicted. In some ways, Terry and Pauline were existing in a temporal register about the here and now and the immediate tomorrow—a marked difference from the tempo of translational research and medicine. The switch in registers was visceral for me. My experiences with them made the *acceleration* at the heart of translational science's expedited logics of research and discovery feel even faster.

## 6.7   Institutions and Academic Livelihoods

The imperative to accelerate science and especially the development of IP therefrom is of course driven by a focus on markets. Yet the politics of academic biomedical research are multidimensional. Despite the revenue the Cleveland Clinic receives for providing clinical services, federal funding has been a key part of its long-standing tradition of research and is vital to its reputation. In fact, even during the financial crisis, which impacted Cleveland quite substantially, federal research funds enabled the organization to create new buildings, initiatives, trials, departments, and positions.

For example, several new programs were initiated and staff hired as part of ELSI funding (Ethical, Legal, Social Implications [of science]) from the National Science Foundation. Not only was this funding central to the creation of these hires and the programs, the symposium that I attended on the risks, benefits, and implications of new brain technologies was also funded by the NSF at about the same time. As a part of an ostensibly more reflective institutional ethical regime now attached to emerging bioscience research (Langlitz 2010; Rabinow and Bennett 2012) and translational research (Chen and Carter 2010; Eyal and Sofaer 2010), federal funding disbursed for science research also often includes funds to study the ethical and social implications thereof.[13]

During my time at the Cleveland Clinic, I began to think about the stakes of translational science for clinician-researchers for whom research and competition are part of professional livelihoods (Lamont 2009). The competition for research dollars has become part and parcel of academic work as these monies are tied to university rankings, departmental productivity, as well as the cultural capital accorded to researchers who bring in research dollars.[14] Yet translational science has also become important for the academic medical enterprises in terms of revenue. Dr. Jones, my neurologist interlocutor, explained this further:

> I think academic [medical] centers, and this is just me thinking about it, but some
> of the money to keep an institution going comes also from developing not only
> pharmaceutical, but also devices that can be used for patients, and that can be
> used to fund more research. They're encouraging us to try to, when we can, create
> a product that can be marketed. Research isn't just for one's curiosity anymore.
> (Interview 12, September 22, 2010)

Thus, the stakes for institutions are large. These developments span an array of academic livelihoods and shape the kind of work that researchers conduct. Translational science includes the proposition of greater research efficiency. Yet the stakes also include the status of the academic institution itself in the West—which, as a result of social and political contexts, is presently in a moment of transition toward a more market-oriented and economically rational site (Etskowitz et al. 2000; Kenner 1986; Strathern 2000). This context has been important for TN—both as a symbol of the stakes that exist for academic medical centers and their constituents and as a potential preview of the future of academic labor and the constitution of academic value. Thus, the crisis facing many academic medical centers— with their unique set of economic challenges—reflects the stakes at work for academic medical centers specifically. In other words, translation reflects a global mandate regarding clinical and academic labor and value (Apple 2005; Olssen and Peters 2005; Orr 1997; Stewart 1997; Strathern 2000) and renders both in market and moral terms through the risks of failing to be "productive." This intermingling of the moral and the market is a hallmark of contemporary biomedical models of innovation (Robinson 2017). Yet for ethicists such as Michael Sandel (2013), this intersection is an inherent problem; it is at the heart of the challenges facing contemporary bioethics. This meeting of markets and medicine, especially in contemporary knowledge economies, will become a primary driver of the emerging realities of global health.

# 7 Conclusion

In the preceding chapters, I sought to unearth several important "realties" of TN and to a larger extent, translational science and medicine. Through forays into conferences, investor meetings, universities, and laboratories, one finds a set of practices, presumptions, actions, and systems that reveal how translation actually "works on the ground." At the same time, adopting a political and economic lens—mapping global economic, financial, and sector-related contexts behind the rise of TSM—unearths a hidden story about translation as an economic configuration, a means of rehabilitating failed, risk-laden pharmaceutical market strategies and delivering shareholder value. Despite being billed as being solely about health or innovation, deeper analyses reveal a more complicated story.

Above and beyond this ethnographic and contextual understanding lies one of the most important questions of all: Does translation actually "deliver" on its grand promises to patients? In addressing a question that is at once empirical and ethical, this book has also sought to explore how the translational shift in neuroscience intersects with the worlds of patients. What are the politics of the translational paradigm in relation to its moral warrant? How are patients both understood and prefigured (or configured) in TN and TSM? Does TSM deliver on its promise to the worlds' patients? In the sections that follow I bring together several arguments made throughout the book and further expand on important questions and implications raised by the translational turn.

## 7.1 Do Translational Science and Medicine Work?

In discussions with interlocutors and academic audiences, I frequently encounter the question about whether translation is "successful." Of course, the question often presumes notions of success that mimic the language of TSM itself. Since its instantiation, few novel medicines or technologies have emerged. For its proponents, TSM is a "long game," meaning that success— new, successful innovative drugs and technologies—may still be a long way off.

Yet this may be the wrong question to ask. If TSM is better analyzed using a political economic lens, a different answer emerges in response to the same question about whether TSM is successful. What one finds is that even where there is no actual "innovation," no improvement in patient health, and even where there has been no novel IP created, something significant has still occurred. For global biopharmaceutical R&D organizations, the externalization of R&D has enabled a divestment of risk, including the transfer of risky research portfolios, a lower overall cost of R&D, and an expanded global pipeline of innovation sources. In this case, *TSM is a fully realized financial outcome*, evidenced by the fact that externalization-focused R&D strategies are listed in pharmaceutical-company shareholder reports as a way to cut R&D costs while retaining access to innovation pipelines. Such strategies deliver returns in the form of shareholder value for the companies that have executed this plan along with other cost- and risk-reduction strategies. Thus, translation already "succeeds" bereft of or prior to any novel innovations, knowledge, effective medications, or improved "health."

Questions subsequently emerge about how to make sense of TSM. Is TSM science? How are we to analyze the increasing financialization of biomedicine and biomedical research? And yet the impulse to think about TSM in solely epistemological terms—about whether its knowledge claims are internally consistent, for example—reflects a growing analytical problem for scholars, especially in analyzing increasingly market-oriented medical models.

## 7.2  Understanding TN Financially

Understanding TN's financial functionality is essential to an empirical understanding of TN and how it works. In this book, I suggest that TSM frames academic science and the university work itself as a set of underleveraged assets. TN-based partnerships essentially function as a component of an early-stage investment strategy on the part of life science investors

and global biopharmaceutical firms. Delving into the concept of externalization, one sees how translational partnerships fit within an increasingly
global strategy. This strategy together with other sectorwide trends such
as R&D downsizing, increasing risk aversion, vertical disintegration, and
a focus on short-term shareholder value help to explain how translation
fits within a particular context. In the specific case of TN, these sector contexts as well as specific challenges particular to the brain sciences help to
explain large-scale divestment from neuroscience-related areas, including
areas such as Alzheimer's and depression, among others.

Through state policy shifts such as the development of translation-
focused programming and funding demands both in the US and globally, one sees the development of a translation-focused infrastructure. Yet
one also sees the NIH hand-select partnerships between pharmaceutical
partners and universities. Between these explicit partnerships and more
informal integration of university translational research programs into a
"discovery pipeline" or network, TN programs became the means through
which university TN programs functioned as R&D arms of biopharmaceutical companies.

Ultimately, I argue that TN—given its capacity to create shareholder value
even where there is little to no novel innovation, as well as its functionality as an early-stage investment strategy—is an enterprise best understood
not as a novel form of science or medicine, but as a form of finance. TSM
also outsources the very riskiest parts of early-stage R&D to universities,
but without the requisite billions in profit that Pharma companies have
enjoyed. At the social level, TSM may also provide a means by which early-
stage research innovations are de-risked via the public sphere, here including public financing by state agencies and execution by nonprofit public
and private universities. It is in these claims where implications abound:
Might emerging models of biomedical innovation constitute a hybrid form
of science and finance, or financial enactments in scientific realms? What
are the politics and costs of increasing public ingestion of private-sector risk
done in the name of "innovation"?

Ultimately, the story of TN is about financialization. The increasing
financialization of the life sciences means that analysts of emerging scientific and biomedical paradigms will need analytics attuned to—or capable of
grasping—issues of finance. At the same time, understanding the financial
functionality of TSM provides clarity regarding the *scientific* particularities

of TN. In other words, understanding TN financially is key to understanding TN epistemologically.

## 7.3  TN's Molecular Starting Points

In chapter 6, I provided a graphic that sought to map the life cycle of translational science and research as well as the translational "continuum" and its constituent parts. What one finds across various graphic representations of TSM is that they tend to depict a molecular starting point. Locating therapeutic targets (biological targets that are or can be manipulated toward a therapeutic effect) becomes the driver and the warrant for translational projects and the way to assess scientific and market value. To be clear, having a molecular starting point enables an easier pathway toward creating a pharmacological solution.

Yet, a second element is revealed. As biopharmaceutical companies externalize their R&D projects and as private industry's claims and desires drive agendas and foci of university-based scientists and entrepreneurs (think here about the Neurotechnology Investing and Partnering Conferences where investors and Pharma executives make proclamations before audiences of university-based scientists and entrepreneurs), one unearths a network of levers between corporate strategies and the birth of scientific projects, experiments, agendas, and questions. These birthplaces are also the space of therapeutic envisioning—where areas of medical focus emerge, trends grow, and decisions about investments in R&D are often made or catalyzed. As TN produces a critical legibility between university science and the market, there are myriad epistemic implications.

## 7.4  Knowledge Constriction in Translational Science and Medicine

From conversations with clinicians and patients to discovering TSM's strange particularities such as its reliance on molecular starting points (shared by TN and other translational fields such as translational oncology), what emerges is an interesting surprise about the nature of translational research. In many ways, translational science and medicine rely on *constrictions* in knowledge—a narrowing of experimental research questions and areas, a defunding of wide swaths of non-application-oriented areas of investigation, a winnowing of therapeutic or pharmacological outcomes, and a marginalization of other potential kinds of clinically meaningful outcomes. TSM may very well lead to asking fewer questions and accepting fewer scientific or medical answers.

Might this winnowing create greater numbers of biomedical objects—solutions, diagnostic tests, and scientific explanations—without a requisite improvement in health outcomes for patients? Perhaps the most obvious evidence of constriction is in translational research's focus on speed. In accelerating the time involved in turning science into products, TSM compresses research life cycles, reducing the time available for greater scientific and clinical understanding. This may potentially create greater numbers of biomedical objects—solutions, diagnostic tests, and scientific explanations—without corresponding improvements in health. In this context where there is knowledge activity that is not necessarily tied to health outcomes, we find evidence of what I call *translational thinking*, part of a set of practices brought about in translational contexts.

### 7.5  "Translational Thinking" and Neuropharmacoepistemology

I use the term *translational thinking* to refer to the ways of thinking in which therapeutic endpoints become the presupposition and warrant for new scientific projects and the means by which the "successes" of translational activity are assessed. The trend especially in universities toward thinking "translationally" is a key result of the translational shift. The impact of translational imperatives on the work of scientists and other researchers requires more extensive investigation.

In the case of TN, the translational obsession is part of a larger phenomenon that I call neuropharmacoepistemology, in which a wide swath of neuroscientific inquiry under the translational imperative is done in the service of specifically pharmacological knowledge aims. The foreclosing of important questions where no pharmacological and biotechnological intervention is immediately envisaged means that wide swaths of research questions must be *reframed* and reimagined for subsumability to a pharmacologically relevant understanding, investigation, and way of knowing. Thus, I use the term to refer both to a mode of thinking and discourse and to the resulting set of logics in which pharmacological objects such as molecular starting points become a necessary scientific grammar for the interrogation of questions in the brain sciences. This mode of thinking also creates questions about a kind of epistemic precariousness introduced through demands placed on neuroscientific thinking and inquiry. How does such a demand contour neuroscience inquiry?

Beyond this epistemic question lie questions about innovations in medicine and health since, of course, pharmacological endpoints or a discovered biomarker, for example—evoking the debate at the Neurotechnology Investing and Partnering Conference,described in chapter 5—are not the same as better health. The trend toward creating "marginally beneficial" drugs—which refers to "new" Pharma-created drugs that involve only minor chemical change and negligible benefits over off-patent drugs—not only shows the impact of financial pressures on drug development, but also, in another sense, the consequence of a neuropharmacoepistemology in which the demand for some marginal, specifically pharmacological novelty becomes "enough" in terms of thinking. This is where neuroscientific imagination and innovation accrete around mere pharmacological relevance.

## 7.6   Partial Subjects and Marginalized Patients

The molecular starting points of TN and TSM; its tendency to originate inquiry around biological objects—genes, glial cells, brains—reflect the fact that under translational inquiry, patients are conjured up in and via their parts. Yet TSM's focus on pharmacological or therapeutic aims not only constructs its objects; it also constructs its *subjects*. As noted, much of my argument in the preceding sections suggests that TSM marginalizes "patients" in favor of people in or via their parts. Translational research relies on what could be thought of as partial subjects, whether via transgenic imaginations at work in the use of humanized mice models or in the elision of patients/people at the outset of the translational continuum. There is a constitution of subjects in or via their parts that lies at the heart of translation. The fact that mice are allowed to "stand in" for humans in experimentation is based on assumptions of similarity or translatability between distinct animal parts (a rat brain, for example) and parts of human subjects. It is in this implication where one can ask if the subsequent marginalization of patient needs that I found in my ethnographic work, and that is also documented elsewhere, is a consequence of TSM's focus on the molecular. In other words, does a science based on only a partial subject help to explain the marginalization of patient voices elsewhere in the translational research pipeline?

And yet, in the world of the clinic, as my interview with a neurologist indicated, questions emerge about the fact that the additional scientific information generated through biomedical science—including facts related to the biological aspects of a given condition—are not the same as improved outcomes and better health. An epistemology of parts still requires a theory as to the relationship between or among those parts and between biology and disease and between disease and symptom. TSM is still a theory of communication, interrelation, and integration.

## 7.7 Fantasies and Commensurability

One of the many challenges that TSM faces is the underlying promise of TSM to translate scientific information into clinical understanding and therapeutics. As a set of systems, architectures, structures, policies, and platforms, TN also emerged out of a set of fantasies about commensurability. Indeed, the very concept of translation obfuscates an even messier truth: translation is composed of myriad subtranslations. To be clear, the translation of science into therapeutics requires bridging a myriad of gaps including those between one experiment and another, across scientific fields, between one laboratory and another, between laboratory science and clinical research, between and across facts, between science and its deployment within a technology, and ostensibly "from experimental therapy to standards of care" (Fischer 2012, 388).

Accordingly, the project of TSM requires vivid fantasies of *commensurability* across disciplines, modalities, objects, and time. While scholars such as Thomas Kuhn and Paul Feyerabend have written about the problem of incommensurability for analyzing and evaluating successive scientific theories (one new theory replacing another), little attention has been paid to the problems that would exist for *integration* across theories, disciplines, methods, and entire fields (bioengineering and neuroscience in the case of brain devices). TSM depends on fictions of translational possibility— narratives that are at constant odds with unruly biologies, problematic animal models, crises of scientific irreproducibility, and so on. What such fantasies reveal is the underlying presumption that scientific knowledge is advanced enough to warrant or support translation. This assumption combined with pressures to accelerate the speed of discovery and development creates a context ripe for unsafe and/or ineffective therapeutics and diagnostics.

## 7.8   How Do Translational Science and Medicine Work on the Ground?

In-depth analyses of several translational sites revealed interesting components of TSM and TN as enacted on the ground. One important aspect is how much the work of translation is shaped by factors external to the scientific laboratory, including events such as the Neurotechnology Investing and Partnering Conference,. These epistemological theaters served as a key space for enacting TN: here I observed the formation of partnerships, alliances, networks, and scientific collaborations. I also witnessed the elucidation of "key" research aims and directions, proclamations about research areas that were "risky" or "opportune," and revelations about where entrepreneurs and academics find private-sector financial backing for ventures. This site was indispensable to understanding TN on the ground, highlighting the importance of closely tracking explicit market spaces where decisions are made about the world.

At the same time, TSM functions through a tapestry of interconnected systems, infrastructures, legal agreements, and technologies. Designs for both software and buildings sought to render translation sustainable, easier, and inevitable. These material environments were instructive because of what they said about how translation was imagined. They were useful artifacts inasmuch as they were also archives of translational fantasies.

## 7.9   Bioethics: Detethering, Financialization, and the Need for an Economic Bioethics

While this book focuses on a single case study of TN, the aim of the larger work is methodological and conceptual, in underscoring the need for a scholarly approach to contemporary models of research and innovation that is attentive to increasingly political and economic dimensions.

I began this foray into TN at the outset of the book via the lens of ethics. It was the need for an ethics of TN on the ground that also inspired this project. In later sections of the book, I describe my attendance at a neuroethics conference, where there was almost no discussion of the ethical issues brought about by the increasingly profit-motivated decisions by Big Pharma that failed to improve health. The reason there was no discussion of these issues (save during the keynote) was because neuroethics, like many fields, focuses largely on epistemology or speculation about the implications of future technologies. Neuroethics attracted scholars interested in applying their skills at logic games to speculative worlds, such as those that might emerge in the future that consists of artificially intelligent robots or neurologically enhanced people. Inattention to issues of political economy is a common challenge for scholarly fields of inquiry that focus on analyzing science and medicine, because many of these fields focus largely on epistemological concerns at the expense of understanding how even these concerns themselves are the product of global economic and financial configurations. Ethics is similarly constrained by its attention to, for example, moral issues between two individuals or decisions between two choices, but without adequate consideration of how a given set of moral choices or moral dilemmas are the product of larger social forces.

If, as I suggest here, TSM's externalizing capacity helps firms lower internal costs and risks from early-stage R&D, and if translational demands may potentially foreclose knowledge pathways, then there is a need to understand the relationship between TN and increasing risks to patients. If TSM leads to economic development but few novel developments for health, then a separate set of ethical questions emerge about the ethics of continued investment in TSM on the part of the public.

In a larger sense, one of the broader moral and social issues is the story of financialization. As I have shown in this book, TN and TSM in general reflect the increasing financialization of innovation in science, medicine, and healthcare. The ethical implications of this shift lie in how TN (and TSM more broadly) effectively *detether* biomedical research efforts from patient health outcomes—a finding that helps shine light on the real, unfolding implications of neuroscience as it moves "from the laboratory to the field" (Abi-Rached 2007). While some ethical considerations regarding TSM correctly point to the problem of potential financial conflicts of interest and their impact on research (Eyal and Sofaer 2010; Maienschein et al. 2008; Sugarman and McKenna 2003), I would suggest an additional ethical issue related to financial interests, namely, the way that TSM requires the subversion (Holman and Elliott 2018; Robinson 2018) of norms of science, health, and medicine, in the service primarily of financial strategies and needs.

In TSM, this detethering occurs at various moments. For example, the focus on creating market-conducive applications from TSM means that translational projects require replacing traditional assessments of academic research—namely, journal publications—with modes of evaluation that "recognize partnerships" with industry. Despite powerful linear innovation narratives, increased IP generation is not the same as improved health. And as TSM seeks to use IP and partnerships as key indicators of progress in lieu of assessments based on patient outcomes (an admittedly difficult kind of indicator), there is a *detethering* effect where applications become the means of assessing the success of translational medicine over actual patient health. As a swath of research about biopharmaceutical progress has shown, the financialization of health and medicine has meant that Pharma companies are able to create profits as well as market and shareholder value without concomitant new innovations in medicines or clear direct impacts on health. This detethering of practices in science and medicine from patient health outcomes is a key ethical implication of the financialization of science and medicine.

The issue of financial entanglements also means that "postapproval" (Eyal and Sofaer 2010) research may not be as free from ethical concerns as some, such as Eyal and Sofaer, suggest. Postapproval research may still reflect (and effectively hide) problematic upstream decision making tied to market imperatives. While postapproval research is certainly less controversial than preapproval research, if financial considerations such as those outlined in this book bring about upstream decisions that winnow the set of solutions that make it through R&D pipelines, or if a focus on short-term shareholder-value concerns causes companies to create innovations that are of marginal patient benefit (Light et al. 2013; Light and Maturo 2015) or are even bereft of patient benefit, then ethically problematic aspects are still buried within postapproval research. In fact, the implication that financial imperatives shape biomedical research across its stages, from laboratory ideation to postapproval research, shows the unusual reach of market pressures on biomedical innovation.

That financialization's ethical stakes are regularly missed (Robinson 2018) indicates the need for bioethicists to consider the market-oriented systems that now inform contemporary ethical dilemmas in science, medicine, and health. That is, the financialization of healthcare means that there is a need for something like an *economic bioethics*. As often-hidden market-driven factors will likely continue to shape new healthcare realities -- from the design of medicines to the emergence of new disease categories—all issues of bioethical concern—there is a concomitant need for a set of ethical tools, approaches, and research models that can deal with the complexity of an increasingly market-oriented system of healthcare innovation and delivery.

# Notes

## Chapter 1

1. http://www.neuroinsights.com/marketreports/marketreport2013.html.

2. CNS (central nervous system) is an acronym that is used to refer to the entire sector of brain-disorder-related research, firms, products, investments, and speculators. Similar to the categorization of cardiac defibrillators under medical devices, "CNS" is especially used by insiders—investors, pharmaceutical and life science companies, and biotechnology companies—to refer to the entire class of research areas and products that surround the brain.

## Chapter 2

1. To be clear, Palin referenced a research project focused on olive fruit flies (from the family *Tephritidae*), which are different from *Drosophila melanogaster*. Nevertheless, the media conversation about her purported disdain for science raged on. Much was made of this incident in blog articles and magazines such as *Scientific American*.

2. For a nuanced analysis of the boundaries between science, risk, and the public—especially around policy debates—see Jasanoff and Kim (2013) as well as Wolfe (2002).

3. here is an excellent overview of the history of the rise of instrumentalism in higher education in North America and its connection to both the Cold War and communist anxiety. Thus, the shift toward instrumentalism has a clear history and reflects clear political shifts on the part of state actors (Grace 2012, 225–244).

4. For more on this, see the 2008 *Guardian* article, "Palin and the Fruit Fly: How the Vice-Presidential Candidate Became a Laughing-Stock among Scientists." To get a sense of the tenor of public sentiment, see http://www.guardian.co.uk/commentis-free/2008/oct/27/sarahpalin-genetics-fruit-flies. Commenters' posts often reference the promises of biomedical research as part of their castigation of Palin's sentiments.

5. "One will also find the term *molecular medicine* used to describe research that seeks to exploit molecular understandings in biomedicine toward greater therapeutic application. In Europe one finds an expansion of the use of the term *molecular medicine*, and many US centers focused on translational science and medicine include both 'Translational Science' or 'Translational Medicine' and 'Molecular Medicine' in their titles or as the title of graduate programs. For example, as of this writing, Baylor College of Medicine in the US has a graduate program in 'Translational Biology and Molecular Medicine.' Boston University is one of several universities with programs in 'Translational and Molecular Medicine.' Centers dedicated to molecular medicine have proliferated in Europe, with many adding 'Translational Medicine' to their name or renaming the institute as 'Translational Medicine.' Given the often-vicissitudinal nature of terminologies and names used for scientific projects (Boniolo and Nathan 2016b), it is important to note that not all name changes have signaled a clear shift in activities. Also, while the translational outputs (in terms of medicines) of TSM and molecular medicine tie the two areas together definitionally, an argument can be made that each field encompasses distinct endeavors and histories that possess significant areas of overlap" (Robinson 2018, 3).

6. http://www.ncats.nih.gov/research/cts/ctsa/funding/apply/tr-12-006.html.

7. National Center for Advancing Translational Sciences. Homepage: https://ncats .nih.gov. It is also important to note that these funds are not newfound funds. In fact, they are the result of a reallocation of existing funds from within the NIH. Thus, this really reflects a shifting set of priorities in terms of how health is being envisioned at least among the stakeholders at NIH. See this interview with Chris Austin, who was appointed director of the National Center for Advancing Translational Sciences in late 2012: http://www.raps.org/focus-online/news/news-article -view/article/3407.aspx.

8. http://officeofbudget.od.nih.gov.

9. http://www.ncats.nih.gov/research/tools/preclinical/patents/tech-licensing.html.

10. The licensing occurs through the Strategic Alliances branch within the Office of Policy, Communications and Strategic Alliances. According to its self-description, it "aims to make it easy for industry and academia to interact and partner with NCATS laboratories and scientists. The branch provides a complete array of services to support NCATS technology development and partnership activities." See http://www .ncats.nih.gov/research/tech-transfer/alliances.html.

11. See http://www.nature.com/neuro/journal/v7/n12/full/nn1204-1281.html.

12. This follows from a surge of research about the commodification of university knowledge (Etzkowitz 2003, 2007; Hong and Walsh 2009; Lam 2010; Strathern 2000), of the general corporatization of the university (Bok 2003; Buchbinder 1993; Shumar 1997), of the role of capitalism in the work of life scientists (Shapin 2008;

Woolgar, Coopmans, and Neyland 2009), along with histories of transformation at the university (Etzkowitz et al. 2000; Lowen 1997). This line of research about the university and industry is not new. Of course, Jean-François Lyotard's (1984) notions about the transformation of knowledge especially for late capitalist society have implications for the nature of knowledge and value.

13. http://www.unmc.edu/pharmacology/integrative_translational_neuroscience .htm.

14. http://www.unmc.edu/pharmacology/integrative_translational_neuroscience .htm.

15. www.nih.gov/about/budget.htm.

16. As I have tracked representations of translational science and translational medicine, I have noted an observable rise in the number of university-based translational center websites that advertise the specific drugs they have developed—as if to indicate the *productivity* of the translational center. I take this focus on the lab's discoveries as intimately linked to a set of anxieties that the translational imperative has been designed to fix.

## Chapter 3

1. Evaluate Pharma. World Preview 2015. 2015. http://info.evaluategroup.com/ rs/607-YGS-364/images/wp15.pdf.

2. The closure of internal R&D departments by large multinational firms was part of a much longer history. US companies that had developed reputations for research and discovery, such as AT&T and Bell Labs, shut down their R&D centers that had focused on early-stage science and technology research and contracted with external firms and found other means of sourcing innovation, including strategies (Mirowski 2011).

3. The very velocity of the university's corporatization and privatization (Raaper and Olssen 2015; Shore and Wright 1999) may necessarily obviate questions about corporate pressures external to the nonprofit university. In other words, the distinction between the nonprofit university and for-profit corporate partners may be an increasingly difficult demarcation.

4. Johnson & Johnson Innovation Centers, n.d.

5. I also don't mean to suggest a one-way linear model of influence in which industry is merely or simplistically apprehending the methods, norms, and on-the-ground realities of academic science. The notion of asymmetrical convergence (Kleinman and Vallas 2001) gets at the many dualities and bidirectionality of industry-university relations. It is also important to avoid ontological distinctions between

"industry" and "universities" that treat each category as distinct and unchanging. Indeed, modern universities increasingly function/appear as corporations in a variety of domains (see Mirowski 2011 especially). Thus, my discussion here references biopharmaceutical-industry stakeholders specifically and their relationship to specific universities and TSM programs. Nevertheless, as I outline in this book, the enveloping of the university and university resources was part of an explicit strategy by private-sector partners and state sponsors.

6. These figures also challenge the argument often offered by critics that the requirement created by legislation such as the Bayh-Dole Act that technology be licensed essentially guarantees public benefit. While Bayh-Dole allows the state to exercise its interest in technologies that are not commercialized, it is not clear that capacity to exercise this option has translated into a stream of innovations that have successfully benefited the public.

7. "Realization" refers to its cultural articulation as a scientific reality rather than some sort of cognitive understanding. The ambiguity of the science behind antidepressants, for example, has existed for a long time.

8. Destigmatization is often pointed to as a desire for the field as well as a good byproduct of the biomedicalization of psychiatric disease. However, destigmatization becomes a glib justification for pharmaceuticalization—it is a common solution that clinicians and pharmaceutical-company representatives harp on, out of what I believe is moral anxiety.

9. It is not clear who Insel meant by "us." It is possible that here he refers to the NIMH, in its early funding of neurobiology research and subsequent dissemination of neuropsychiatric models, especially via diagnostic categories (Lee, Ng, and Lee 2008).

10. This refers to the notion of geneticization that epidemiologist Abby Lippman (1992) offered in order to make sense of the ways genetics became a driving force in medical meaning making. For Lippman, geneticization was a form of reduction/ simplification.

11. The constant use of the term *bridge* in translational programs, research articles, reports, and communication shows both its conceptual tie to "connection" and the gulfs that exist in the translational imaginary created for neuroscience.

12. Ironically, Kaushik Sunder Rajan's (2006, 1) ethnography also included an interlocutor who had read Paul Rabinow and who was familiar with anthropologists working among scientists.

13. This also sets the therapeutic bar very low: A "quarter-assed" (Grosof) effect would yield a positive financial outcome.

14. See http://www.nimh.nih.gov/health/publications/the-numbers-count-mental -disorders-in-america/index.shtml.

15. The particular relevance of neoliberalism in thinking about TN is in the way it fits with the rhetoric of defunding the public university, in this case seen as the reallocation of federal funding to market-oriented research and not basic science, and perhaps as the general defunding trends regarding the public university (Mirowski 2011) and federal research dollars. Neoliberalism is especially germane in its forwarding of privatization interests. Thus, as TN creates new opportunities for public-private partnerships, it is consistent with this element of neoliberal ideologies and represents a good case study for the scientific manifestation of neoliberalism's ideology. However, there is much more to TN than its neoliberal legibility.

16. In this sense, perhaps neoliberalism at the university had been appropriated in the interests of other social and technological dramas (Pfaffenberger 1992).

17. This is partly the reason I focused on TN rather than other translational disciplines, such as translational biology.

## Chapter 4

1. "Biotechnology Payroll Tax Exclusion" (n.d.), http://oewd.org/Biotechnology -Payroll-Tax-Exclusion.aspx.

2. Bayer, "The CoLaborator, an Incubator Space for Start-Ups," n.d., http://www .colaborator.bayer.com/en/about-the-colaborator/.

3. In addition to liquefaction, the history of industrial contamination at Mission Bay (now rendered nearly invisible because of the area's rapid transformation) left toxins in the soil. Built near former railyards, toxicity levels were sufficient to spur a petition to move a UCSF daycare center away from Mission Bay.

4. The terms of the agreements were something that I wanted to explore further; however, no stakeholder or entity was willing to share these agreements or even talk about them.

5. UCSF's CTSI is not to be confused with Pfizer's CTI that I mentioned earlier.

6. As I explored the system, I noticed how similar the small logo atop the UCSF Profiles site (in figures 4.7 and 4.8, resembling a small human head connected to a host of other heads) was to the network image found at AstraZeneca Neuroscience's iMed website (a rainbow of small circles connected by sharp lines into a network anchored by an AstraZeneca logo) (figure 4.5).

7. For Latour, this displacement also undergirds the mystification and black boxing of science.

8. Translational science, the commercialization of science, and the growth of public-private partnerships at the university present a useful analytical entry point for science and technology scholars interested in following a fact through its translation.

Scholars such as Bryan Wynne (1993) pioneered empirical research about how facts moved outside of the laboratory into the world, and the role various publics in science have in this process (a concern the British took up earlier than their American counterparts due to the structural particularities of British universities). In *Misunderstanding Science? The Public Reconstruction of Science and Technology*, Wynne and Wynne's (1996) project of getting at public understanding necessarily required a cognitive consideration. The cognitive methods in their research on Cumbrian farmers and scientists in a post-Chernobyl context compelled a different mode of understanding the fundamental translation project. In thinking of TN as an institutionalized assimilation of new knowledge by existing knowledge, my project could be one about thinking. More than a study of scientific production, mine is a study of knowledge representation and, to some extent, learning. Yet, part of my larger proposal is that understanding or misunderstanding is not central to TN work to the extent to which poor or problematic scientific understanding is not necessarily an impediment to translational "success." Because TN is also about commercialization, a cognitivist approach toward the study of TN would miss how TN is a much more complex phenomenon. Mistranslation is a necessary and intended part of the translation process rendered less important by the engines of the market and speculative investing in biomedicine. This is why I did not take a cognitivist approach in this project.

9. If classic accounts of scientists' work sounded like objective, interest-free pursuits of truth, or "pure science," the translational model of science was instrumentalists doing science for a clear goal. Scholars such as Sandra Harding (1993), Donna Haraway (1997), and Sharon Traweek (1988) have thoroughly reflected on and intervened in these narratives and shown their inextricability from the work of gender and in the construction of "cultures of objectivity" that marked narratives of scientific practice (Rosenberg 1997). Rather, this highly interested science wore its values on its sleeve. This was not the classic, "objective" "disinterested science," that were the paragons of historical accounts of scientific discovery. Historian Steven Shapin (2008) delineated this classic scientific mode as one characterized by an honor among scientists tied to one's independence toward a new paradigm marked by experimentalism.

10. I take up the role of imagination and fantasy in translational research in a separate publication, asking about fantasies of commensurability between and across disciplinary fields—an issue of early concern for the philosophy of science as well as other fields dealing with knowledge commensurability.

11. See Collins (1999) for research on the phenomenon of lab workers putting up obstructions to obscure glass in glass-oriented laboratory architecture.

12. It is important to note that there was a great variety of architectural opinion regarding laboratory design, as was the case with design in general. For some, openness and flexibility should be seen less as an attempt at engineering social relations

and more as an environment tooled for conductivity and "accommodation," as architect Robert Venturi (1999) contended in *"Thoughts on the Architecture of the Scientific Workplace: Community, Change, and Continuity."* Also see Solomon (1993).

13. Why were relationships necessary? One thing I learned was that knowledge did not translate itself. Often laboratory results were so difficult to reproduce by companies that owned the patent for a given piece of research that increasingly, part of the very patent included (necessarily) consulting by the scientist that created the result in the initial laboratory experiment. This has shown that the actual transfer process included arranging for a legally sanctioned series of social relations—to confer what legal scholar Peter Lee (2012) has called embodied knowledge, which is necessarily part of scientific research.

14. In his essay "Bridge and Door" (1909), Simmel actually discussed the nature of the window and the meaning of its transparency. However, he suggested that the window, unlike the bridge, was more often used by the insider gazing out and thus conferred on it only "part of the significance of the door." In a world of glass walls and buildings with glass exteriors, I propose that Simmel's door functions as a better metaphor—especially in thinking about how the increasingly "open" laboratory functions as a kind of doorway to new scientific knowledge and practice.

## Chapter 5

1. There is also an opportunity to think about materiality via financialization. In other words, how might financial relevance become the means through which things "matter"—objects that are produced or that emerge via becoming, or by virtue of being connected to, a financial "matter"? What ecosystems of health technologies are "birthed" through the demands of financialization on markets, scientists, entrepreneurs, and other stakeholders that bring about new medicines, concepts, and solutions? And, following Stefan Helmreich's (2018) provocation, a question emerges about other ways in which the biological and the economic are commingled.

2. One also finds this fusing together in the growth of professional master's programs that are now being created to produce business-minded scientists as well as science-versant business executives.

3. This is especially the case among venture capital communities. There is interesting scholarship (Sharp 2013) about the role of social networks and social capital in the machinations of investment and information-gathering processes, an issue that has been well documented (Cohen and Fields 2000; Dierkes and Hoffmann 1992; Kenney and Florida 2000; Saxenian 1996; Welz 2003).

4. There is resonance in thinking about parts in discussions of the fragmentation of bodies as elucidated in the work of scholars such as Leslie Sharp, Margaret Lock, and

Judith Farquhar, which they further articulate in their discussions of materiality, the commodification of the body and its parts (Sharp 2000), as well as embodiment. Yet there is also an opportunity to connect it to the way it produces partial subjects when such parts come to stand in for "the whole" of subjects. It is also important to think about the ethical and epistemological aspects of these forms of embodiment.

## Chapter 6

1. National Survey of Children's Health. NSCH 2007. Data query from the Child and Adolescent Health Measurement Initiative, Data Resource Center for Child and Adolescent Health website, www.childhealthdata.org.

2. For more information on the disparity between the Cleveland Clinic's wealth and that of the city, see Herper (2012).

3. http://quickfacts.census.gov/qfd/states/39/3916000.html.

4. See "Deep Brain Stimulation Reduces Epileptic Seizures in Patients with Refractory Partial and Secondarily Generalized Seizures," *ScienceDaily,* March 18, 2010, http://www.sciencedaily.com/releases/2010/03/100318082016.htm, and also "More Good News for Deep Brain Stimulation in Epilepsy," *Medscape,* December 3, 2012.

5. In fact, the history of DBS as a method emerges from an accidental discovery in 1870 that stimulation in animals was seen to have an unexpected clinical use possibility (Kringelbach et al. 2007).

6. Also, it is important to note that all clinicians may not share the sense of urgency that the FDA applied to the recall. There were clinicians who disagreed with the alacrity of the FDA recall. For more, see Walsh (2013b).

7. http://www.bloomberg.com/news/2013-05-02/medtronic-deep-brain-stimulator -cited-by-fda-for-flaws.html.

8. http://www.fda.gov/AboutFDA/CentersOffices/OfficeofMedicalProductsand Tobacco/OfficeofScienceandHealthCoordination/ucm2018190.htm.

9. Not only is the regulation controversial, but it's also subject to what is arguably misuse by biopharmaceutical companies. Critics (Fins et al. 2011) point to the fact that data gathered through this use can still be used for the marketing of the intervention, during a second appeal to the FDA for approval, or for marketing of the product in different countries.

10. This production of epistemic risk is my principal argument about TN: TN may produce ontological and epistemic risk, precisely by virtue of the fact that it closes down knowledge avenues and also speeds up discovery cycles. In the way that TN acts specifically on knowledge pathways, it produces a specific kind of risk. This issue of the risks engendered into knowledge forms and their impacts has been a

focus of scholars' thinking about the role of data and specific kinds of evidence in global health (Adams 2013; Biehl and Petryna 2013; Fischer 2013).

11. For a more specific example, Medtronic, the biotechnology company that developed DBS devices, had aggressively tried to rebrand DBS devices as an option for epilepsy (MacDonald 2011) and increasingly for psychiatric disorders—a move that garnered controversy at the time. However, while pushing "off-label" use of existing compounds and drugs is a common revenue strategy for Big Pharma (Kesselheim, Mello, and Studdert 2011), the promotion of off-label use also represents a context in which large gaps in knowledge (such as the effects of a drug used for a disorder for which it was not tested) are folded into new biotechnological solutions and produce enormous new patient risks (Dresser and Frader 2009). A social constructivist (Boholm 2003) and/or denaturalized view of risk provides a useful way to view not only increased risks from potentially premature biotechnological development, but also risks that emerge from the kinds of exceptions provided to experimental biotechnologies via Humanitarian Device Exemptions. It suggests that creating biotechnologies based on problematic knowledge *is already* a social decision about risk. Here we can think of this social tableau as one that is composed of bioengineers, investors, patients, clinicians, and regulators—all implicated in an ethos in which biotechnologies get understood as the most important means of intervening in health.

12. In a case of digital irony, my word processing software autocorrects *constrictions* and turns it into *constructions*.

13. This work was not funded directly by these funds. Also, it is important to note that scholars have often pointed out how bioethics is impacted (problematically) by its institutionalization within medical schools and by the sources from which it has received funding. However, all federal research possesses this complication.

14. Research grants are still important for universities' reputations and productivity despite evidence that the costs of research are increasingly falling to the university itself. (See Ehrenberg et al. 2003.)

# References

Aarden, Erik, Alessandro Blasimme, Dustin Holloway, and Luca Marelli. 2015. "Making Sense of Clinical Translation: Ethical, Regulatory and Policy Challenges for Europe and the US." Working Paper from *Making Sense of Clinical Translation: Ethical, Regulatory & Policy Challenges,* Geneva, Switzerland.

Abbott, Allison. 2011. "Novartis to Shut Brain Research Facility." *Nature,* 480, no. 7376: 161–162.

Abi-Rached, Joelle M. 2007. "Neuroscience and Society: A Multidendritic Neuron." BIOS (Centre for the Study of Bioscience, Biomedicine, Biotechnology and Society), London School of Economics and Political Science, Brain Self and Society Paper No. 1. http://www.lse.ac.uk/collections/brainSelfSociety/.

Abi-Rached, Joelle M. 2008a. "The Implications of the New Brain Sciences." *European Molecular Biology Organization Reports,* 9, no. 12: 1158–1162. http://dx.doi.org/10.1038/embor.2008.211.

Abi-Rached, Joelle M. 2008b. "The New Brain Sciences: Field or Fields?" BIOS (Centre for the Study of Bioscience, Biomedicine, Biotechnology and Society), London School of Economics and Political Science, Brain Self and Society Paper No. 2. http://www.academia.edu/236035/ The_New_Brain_Sciences_Field_or_Fields.

Adams, Vincanne. 2013. "Evidence-Based Global Public Health: Subjects, Profits, Erasures." In *When People Come First: Critical Studies in Global Health,* ed. João Biehl and Adriana Petryna, 54–90. Princeton, NJ: Princeton University Press. http://www.jstor.org/stable/j.ctt2jc895.

Adams, Vincanne, ed. 2016. *Metrics: What Counts in Global Health.* Durham, NC: Duke University Press Books.

Adams, Vincanne, and Sharon R. Kaufman. 2011. "Ethnography and the Making of Modern Health Professionals." *Culture, Medicine, and Psychiatry,* 35, no. 2: 313–320. http://dx.doi.org/10.1007/s11013-011-9216-0.

Addison, Courtney. 2017. "Bench, Bedside, Boardroom: Negotiating Translational Gene Therapy." *New Genetics and Society* 36, no. 1: 22–42. https://doi.org/10.1080/14636778.2017.1289468.

Agamben, Giorgio, and Kevin Attell. 2005. *State of Exception*. Chicago: University of Chicago Press.

Allday, Erin. 2013. "UCSF Mission Bay Marks 10 Years." *SFGate*, January 23. http://www.sfgate.com/bayarea/article/UCSF-Mission-Bay-marks-10-years-4215668.php#src=fb.

Alpert, Susan. 2011. "A Humanitarian Device Exemption for Deep Brain Stimulation: Author Reply." *Health Affairs*, 30, no. 6: 1212. http://dx.doi.org/10.1377/hlthaff.2011.0425.

Althusser, Louis. 1971. "Ideology and Ideological State Apparatuses." In *Lenin and Philosophy and Other Essays*, ed. Louis Althusser, 127–188. New York: Monthly Review Press.

Andersson, Tord, Pauline Gleadle, Colin Haslam, and Nick Tsitsianis. 2010. "Bio-Pharma: A Financialized Business Model." In "Critical Perspectives on Taxation," ed. Rebecca Boden, Sheila Killian, Emer Mulligan, and Lynne Oats, special issue, *Critical Perspectives on Accounting*, 21, no. 7: 631–641. https://doi.org/10.1016/j.cpa.2010.06.006.

Appadurai, Ariun. 1988. *The Social Life of Things*. Cambridge: Cambridge University Press.

Applbaum, Kalman. 2006. "Educating for Global Mental Health." In *Global Pharmaceuticals: Ethics, Markets, Practices*, ed. Adriana Petryna, Andrew Lakoff, and Arthur Kleinman, 85–110. Durham, NC: Duke University Press.

Applbaum, Kalman. 2009a. "'Consumers are Patients!' Shared Decision-Making and Treatment Non-Compliance as Business Opportunity." *Transcultural Psychiatry*, 46, no. 1: 107–130. http://dx.doi.org/10.1080/13648470.2010.493707.

Applbaum, Kalman. 2009b. "Getting to Yes: Corporate Power and the Creation of a Psychopharmaceutical Blockbuster." *Culture, Medicine and Psychiatry*, 33, no. 2: 185–215. http://dx.doi.org/10.1007/s11013-009-9129-3.

Applbaum, Kalman. 2009c. "Is Marketing the Enemy of Pharmaceutical Innovation?" *Hastings Center Report*, 39, no. 4: 13–17. http://www.academia.edu/226562/Is_Marketing_the_Enemy_of_Pharmaceutical_Innovation.

Apple, Michael W. 2005. "Education, Markets, and an Audit Culture." *Critical Quarterly*, 47, no. 1–2: 11–29.

Atkinson-Grosjean, Janet. 2006. *Public Science, Private Interests: Culture and Commerce in Canada's Networks of Centres of Excellence.* Toronto: University of Toronto Press, Scholarly Publishing Division.

Avorn, Jerry, and Aaron S. Kesselheim. 2011. "The NIH Translational Research Center Might Trade Public Risk for Private Reward." *Nature Medicine,* 17 (October): 1176.

Baker, Monya. 2016. "1,500 Scientists Lift the Lid on Reproducibility." *Nature News,* 533, no. 7604: 452. https://doi.org/10.1038/533452a.

Baldas, Vassilis, Charalampos Lampiris, Christos N. Capsalis, and Dimitros Koutsouris. 2011. "Early Diagnosis of Alzheimer's Type Dementia Using Continuous Speech Recognition." In *Wireless Mobile Communication and Healthcare,* ed. James C. Lin and Konstantina S. Nikita, 105–110. Berlin: Springer.

Barad, Karen. 2007. *Meeting the Universe Halfway: Quantum Physics and the Entanglement of Matter and Meaning.* Durham, NC: Duke University Press.

Barben, Daniel. 1998. "The Political Economy of Genetic Engineering: The Neoliberal Formation of the Biotechnology Industry." *Organization & Environment,* 11, no. 4: 406–420. https://doi.org/10.1177/0921810698114004.

Barbrook, Richard, and Andy Cameron. 1996. "The California Ideology." *Science as Culture,* 6, no. 1: 44–72.

Basken, Paul. 2011. "Public Research Universities Get Advice from Industry: Please Your Patrons." *Chronicle of Higher Education,* February 20, Government sec. http://chronicle.com/article/Public-Research-Universities/126464/.

Bateson, Gregory. 2000. *Steps to an Ecology of Mind: Collected Essays in Anthropology, Psychiatry, Evolution, and Epistemology.* Chicago: University of Chicago Press.

Baudrillard, Jean. 2005. *The System of Objects.* Trans. James Benedict. London: Verso. (Orig. pub. 1965.)

Bayer, n.d. "The CoLaborator, an Incubator Space for Start-Ups," n.d., http://www.colaborator.bayer.com/en/about-the-colaborator.

Beaulieu, Anne. 2001. "Voxels in the Brain: Neuroscience, Informatics and Changing Notions of Objectivity." *Social Studies of Science,* 31, no. 5: 635–680. http://virtualknowledgestudio.nl/documents/_annebeaulieu/voxels-in-the-brain.pdf.

Beaulieu, Anne. 2002. "A Space for Measuring Mind and Brain: Interdisciplinarity and Digital Tools in the Development of Brain Mapping and Functional Imaging, 1980–1990." *Brain and Cognition,* 49, no. 1: 13–33. http://dx.doi.org/10.1006/brcg.2001.1461.

Beaulieu, Anne. 2003. "Brains, Maps and the New Territory of Psychology." *Theory & Psychology*, 13, no. 4: 561–568. http://www.virtualknowledgestudio.nl/staff/anne -beaulieu/brainsmapsterritory.pdf.

Beck, Ulrich. 1992. *Risk Society: Towards a New Modernity*. London: Sage.

Bell, Vaughn. 2013. "Changing Brains: Why Neuroscience Is Ending the Prozac Era." *The Guardian*, September 22. http://www.theguardian.com/science/2013/sep/22/ brains-neuroscience-prozac-psychiatric-drugs.

Belluck, Pam. 2011. "Updated Guidelines for Diagnosing Alzheimer's." *New York Times*, April 19. http://www.nytimes.com/ 2011/04/19/health/19alzheimer.html ?_r=1&src=recg.

Benjamin, Ruha. 2014. "Race for Cures: Rethinking the Racial Logics of 'Trust' in Bio-medicine." *Sociology Compass, 8, no.* 6, 755–769. https://doi.org/10.1111/soc4.12167.

Benjamin, Walter. 1968. "The Task of the Translator: An Introduction to the Trans-lation of Baudelaire's Tableaux Parisiens." In *Illuminations: Essays and Reflections*, ed. Hannah Arendt, 69–82. New York: Schocken Books. (Orig. pub. 1923.)

Berman, Elizabeth Papp. 2012. *Creating the Market University: How Academic Science Became an Economic Engine*. Princeton, NJ: Princeton University Press.

Biehl, João. 2005. *Vita: Life in a Zone of Social Abandonment*. Berkeley: University of California Press.

Biehl, João. 2007. *Will to Live: AIDS Therapies and the Politics of Survival*. Princeton, NJ: Princeton University Press.

Biehl, João. 2013. "The Judicialization of Biopolitics: Claiming the Right to Phar-maceuticals in Brazilian Courts." *American Ethnologist*, 40, no. 3: 419–436. http:// dx.doi.org/10.1111/amet.12030.

Biehl, João, Byron J. Good, and Arthur Kleinman. 2007. *Subjectivity: Ethnographic Investigations*. Berkeley: University of California Press.

Biehl, João, and Peter Locke. 2010. "Deleuze and the Anthropology of Becoming." *Current Anthropology*, 51, no. 3: 317–351. http://www.jstor.org/stable/10.1086/ 651466.

Biehl, João, and Adriana Petryna. 2011. "Bodies of Rights and Therapeutic Markets." *Social Research*, 78, no. 2: 359–394.

Biehl, João, and Adriana Petryna. 2013. "Therapeutic Markets and the Judicialization of the Right to Health." In *When People Come First: Critical Studies in Global Health*, ed. João Biehl and Adriana Petryna, 325–346. Princeton, NJ: Princeton University Press.

Bijker, Wiebbe E. 1997. *Of Bicycles, Bakelites, and Bulbs: Toward a Theory of Sociotechnical Change*. Cambridge, MA: MIT Press.

Birch, Kean. 2006. "The Neoliberal Underpinnings of the Bioeconomy: The Ideological Discourses and Practices of Economic Competitiveness." *Genomics, Society and Policy*, 2, no. 3: 1–15. https://doi.org/10.1186/1746-5354-2-3-1.

Birch, Kean. 2017a. "Rethinking *Value* in the Bio-Economy: Finance, Assetization, and the Management of Value." *Science, Technology & Human Values*, 42, no. 3: 460–490. https://doi.org/10.1177/0162243916661633.

Birch, Kean. 2017b. "Techno-Economic Assumptions." *Science as Culture*, 26, no. 4: 433–444.

Birch, Kean, and David Tyfield. 2013. "Theorizing the Bioeconomy: Biovalue, Biocapital, Bioeconomics or . . . What?" Science, Technology & Human Values, 38, no. 3: 299–327. http://dx.doi.org/10.1177/0162243912442398.

Birch, Kean. 2017c. "Financing Technoscience: Finance, Assetization and Rentiership." In The Routledge Handbook of the Political Economy of Science, edited by David Tyfield, Rebecca Lave, Samuel Randalls, and Charles Thorpe. Oxford: Routledge.

Black, Max. 1990. Perplexities: Rational Choice, the Prisoner's Dilemma, Metaphor, Poetic Ambiguity, and Other Puzzles. Ithaca, NY: Cornell University Press.

Bloch, Maurice. 2012. Anthropology and the Cognitive Challenge. Cambridge: Cambridge University Press.

Boellstorff, Tom. 2008. Coming of Age in Second Life: An Anthropologist Explores the Virtually Human. Princeton, NJ: Princeton University Press.

Boholm, Åsa. 2003. "The Cultural Nature of Risk: Can There Be an Anthropology of Uncertainty?" Ethnos, 68, no. 2: 159–178.

Bok, Derek. 2003. *Universities in the Marketplace: The Commercialization of Higher Education*. Princeton, NJ: Princeton University Press.

Bole, Kristen. 2011. "UCSF Receives $112 Million to Help Translate Science into Cures." University of California–San Francisco, July 18. http://www.ucsf.edu/news/2011/07/10258/ucsf-receives-112-million-help-translate-science-cure.

Boniolo, Giovanni, and Marco J. Nathan, eds. 2016a. *Philosophy of Molecular Medicine: Foundational Issues in Research and Practice*. New York: Routledge.

Boniolo, Giovanni, and Marco J. Nathan. 2016b. "Philosophy of Molecular Medicine: Introduction." In *Philosophy of Molecular Medicine: Foundational Issues in Research and Practice*, ed. Giovanni Boniolo and Marco J. Nathan, 1–12. New York: Routledge.

Bowker, Geoffrey C., and Susan Leigh Star. 1999. *Sorting Things Out: Classification and Its Consequences*. Cambridge, MA: MIT Press.

Brett, A. 1945. "Instrumentalism and the Humanities." *Bulletin of the American Association of University Professors*, 31, no. 2: 181–188.

Broad, William J. 2014. "Billionaires with Big Ideas Are Privatizing American Science." *New York Times*, March 15. http://www.nytimes.com/2014/03/16/science/billionaires-with-big-ideas-are-privatizing-american-science.html.

Brown, James Robert. 2000. "Privatizing the University: The New Tragedy of the Commons." *Science*, 290, no. 5497: 1701–1702. http://dx.doi.org/10.1126/science.290.5497.170.

Brown, Nik. 2003. "Hope against Hype: Accountability in Biopasts, Presents and Futures." *Social Studies of Science*, 16, no. 2: 3–21. http://www.sciencetechnologystudies.org/v16n2/BrownPDF.

Buchbinder, Howard. 1993. "The Market Oriented University and the Changing Role of Knowledge." *Higher Education*, 26, no. 3: 331–347.

BusinessWire. 2009. "San Francisco at Leading Edge of Global Neurotechnology Industry, Says New Report." *Business Wire*, October 16. http://www.businesswire.com/news/home/20091016005777/en/San-Francisco-Leading-Edge-Global-Neurotechnology-Industry.

Callon, Michel. 1986. "Some Elements of a Sociology of Translation: Domestication of the Scallops and the Fishermen of St Brieuc Bay." In *Advances in Social Theory and Methodology: Toward an Integration of Micro- and Macro-Sociologies*, ed. Karin Knorr-Cetina and A. V. Cicourel, 196–223. New York: Routledge.

Cambrosio, Alberto, Peter Keating, Simon Mercier, Grant Lewison, and Andrei Mogoutov. 2006. "Mapping the Emergence and Development of Translational Cancer Research." *European Journal of Cancer (Oxford, England: 1990)* 42 no. 18: 3140–3148. https://doi.org/10.1016/j.ejca.2006.07.020.

Canguilhem, Georges. 1978. *On the Normal and the Pathological*. Dordrecht, NL: Reidel.

"CBID Establishes Technology Accelerator Fund." n.d. *Projects*. Johns Hopkins University. http://eng.jhu.edu/wse/cbid/page/cbid-establishes-technology-accelerator-fund.

Changeux, Jean-Pierre, and Paul Ricoeur. 2002. *What Makes Us Think? A Neuroscientist and a Philosopher Argue about Ethics, Human Nature, and the Brain*. Princeton, NJ: Princeton University Press.

Chen, Jiin-Yut, and Michele Carter. 2010. "Bioethics and Post-Approval Research in Translational Science." *American Journal of Bioethics*, 10, no. 8: 35–37. http://dx.doi.org/10.1080/15265161.2010.494220.

Choudhury, Suparna, and Jan Slaby, eds. 2012. *Critical Neuroscience: A Handbook of the Social and Cultural Contexts of Neuroscience*. Chichester: Wiley-Blackwell.

Slaby, Jan, and Suparna Choudhury. 2018. "Proposal for a Critical Neuroscience." In *The Palgrave Handbook of Biology and Society*, edited by Maurizio Meloni, John Cromby, Des Fitzgerald, and Stephanie Lloyd, 341–370. Springer.

Chubb, Ian. 2012. "Can Australia Afford to Fund Translational Research?" Keynote Address presented at the Biomelbourne Network, Australia, April 3. http://www.chiefscientist.gov.au/2012/04/can-australia-afford-to-fund-translational-research/.

Clarke, Adele E. 2005. *Situational Analysis: Grounded Theory After the Postmodern Turn*. Thousand Oaks, CA: Sage.

Clarke, Adele E., Janet K. Shim, Laura Mamo, Jennifer Ruth Fosket, and Jennifer R. Fishman. 2003. "Biomedicalization: Technoscientific Transformations of Health, Illness and US Biomedicine." *American Sociological Review*, 68, no. 2: 191–194. http://www.jstor.org/stable/1519765.

Clinical and Translational Science Awards. n.d. "About the CTSA Consortium." https://www.ctsacentral.org/about-us/ctsa.

Cohen, Stephen S., and Gary Fields. 2000. "Social Capital and Capital Gains: An Examination of Social Capital in Silicon Valley." In *Understanding Silicon Valley: The Anatomy of an Entrepreneurial Region*, ed. Martin Kenney, 191–194. Stanford, CA: Stanford University Press.

Collier, Stephen J. 2005. "Global Assemblages." In *Global Assemblages: Anthropological Problems*, ed. Aihwa Ong and Stephen J. Collier, 3–21. Malden, MA: Blackwell.

Collins, James Jr. 1999. "The Design Process for the Human Workplace." In *The Architecture of Science*, ed. Peter Galison and Emily Thompson, 399–412. Cambridge, MA: MIT Press.

Conor, Casey. 2007. "San Francisco's 'Butchertown' in the 1920s and 1930s: A Neighborhood Social History." In *The Argonaut: Journal of the San Francisco Historical Society*, 18, no. 1: 66–83.

Contopoulos-Ioannidis, Despina G., Evangelia E. Ntzani, and John P. A. Ioannidis. 2003. "Translation of Highly Promising Basic Science Research into Clinical Applications." *American Journal of Medicine*, 114, no. 6: 477–484. doi:10.1016/S0002-9343(03)00013-5.

Cooper, Melinda. 2008. *Life as Surplus: Biotechnology and Capitalism in the Neoliberal Era*. Seattle: University of Washington Press.

Cortez, Michele Fay. 2013. "Medtronic Deep Brain Stimulator Cited by FDA for Flaws." *Bloomberg*. http://www.bloomberg.com/news/2013-05-02/medtronic-deep-brain-stimulator-cited-by-fda-for-flaws.html.

Cressey, Daniel. 2011. "Psychopharmacology in Crisis." *Nature News*, June 14. http://dx.doi.org/10.38/news.2011.367.

Crick, Malcolm R. 1982. "Anthropology of Knowledge." *Annual Review of Anthropology*, 11, no. 1: 287–313. http://dx.doi.org/10.1146/ annurev.an.11.100182.001443.

Cutler, Neal R., and Henry J. Riordan. 2011. "The Death of CNS Drug Development: Overstatement or Omen?" *Journal for Clinical Studies* 3, no. 6: 12–15. http://issuu .com/mark123/docs/jcs-volume3-issue6.

D'Andrade, Roy G. 1995. *The Development of Cognitive Anthropology*. Cambridge: Cambridge University Press.

D'Andrade, Roy G. 2001. "A Cognitivist's View of the Units Debate in Cultural Anthropology." *Cross-Cultural Research*, 35, no. 2: 242–257. http://dx.doi.org /10.1177/106939710103500208.

Dai, Ke-Rong, Fei Yang, and Yao-Kai Gan. 2013. "Development of Translational Medicine in China: Foam or Feast?" *Journal of Orthopaedic Translation* 1 (1): 6–10. https://doi.org/10.1016/j.jot.2013.07.003.

Das, Veena, and Ravendra Das. 2006. "Pharmaceuticals in Global Ecologies: Register of the Local." In *Global Pharmaceuticals*, ed. Adriana Petryna, Andrew Lakoff, and Arthur Kleinman, 171–205. Durham, NC: Duke University Press.

Davies, Gail. 2012. "What Is a Humanized Mouse? Remaking the Species and Spaces of Translational Medicine." *Body & Society*, 18, no. 3–4: 126–155. http://dx.doi.org/ 10.1177/1357034X12446378.

Davis, Elizabeth Anne. 2010. "The Antisocial Profile: Deception and Intimacy in Greek Psychiatry." *Cultural Anthropology*, 25, no. 1: 130–164. http://dx.doi.org/ 10.1111/j.1548-1360.2009.01054.x.

Davis, Nikola. 2017. "Over Half of New Cancer Drugs 'Show No Benefits' for Survival or Wellbeing." *The Guardian*, October 5. https://www.theguardian.com/business/2017/ oct/05/over-half-of-new-cancer-drugs-show-no-benefits-for-survival-or-wellbeing.

Dierkes, Meinolf, and Ute Hoffmann. 1992. "Understanding Technological Development as a Social Process." In *New Technology at the Outset: Social Forces in the Shaping of Technological Innovations*, ed. Meinolf Dierkes and Ute Hoffmann, 9–13. Frankfurt: Campus Verlag.

Doganova, Liliana, and Fabian Muniesa. 2015. "Capitalization Devices: Business Models and the Renewal of Markets." In *Making Things Valuable*, ed. Martin Kornberger, Lise Justesen, Anders Koed Madsen, and Jan Mouritsen. Oxford: Oxford University Press.

Douglas, Mary. 1973. *Rules and Meanings: The Anthropology of Everyday Knowledge*. Abingdon, UK: Routledge.

Douglas, Mary, and Aaron Wildavsky. 1983. *Risk and Culture: An Essay on the Selection of Technological and Environmental Dangers*. Berkeley: University of California Press.

Dresser, Rebecca, and Joel Frader. 2009. "Off-Label Prescribing: A Call for Heightened Professional and Government Oversight." *Journal of Law, Medicine & Ethics: A Journal of the American Society of Law, Medicine & Ethics*, 37, no. 3: 476–486. doi :10.1111/j.1748-720X.2009.00408.x.

Dumit, Joseph. 2000. "When Explanations Rest: 'Good-Enough' Brain Science and the New Socio-Medical Disorders." In *Living and Working with New Medical Technologies: Intersections of Inquiry*, vol. 8, ed. Margaret Lock, Allan Young, and Arturo Cambrosio, 209–232. Cambridge: Cambridge University Press. http://dx.doi.org/10.1017/CBO9780511621765.010.

Dumit, Joseph. 2003. "Is It Me or My Brain? Depression and Neuroscientific Facts." *Journal of Medical Humanities*, 24, no. 1: 35–47. http://dx.doi.org/ 10.1023/A:1021353631347.

Dumit, Joseph. 2004. *Picturing Personhood: Brain Scans and Biomedical Identity*. Princeton: Princeton University Press.

Dumit, Joseph. 2012a. *Drugs for Life: How Pharmaceutical Companies Define Our Health*. Durham, NC: Duke University Press.

Dumit, Joseph. 2012b. "Prescription Maximization and the Accumulation of Surplus Health in the Pharmaceutical Industry: The BioMarx Experiment." In *Lively Capital: Biotechnologies, Ethics, and Governance in Global Markets*, ed. Kaushik Sunder Rajan, 45–92. Durham, NC: Duke University Press.

Dumit, Joseph, and Nathan Greenslit. 2005. "Medical Communication with Patients." In *Science, Technology, and Society: An Encyclopedia*, ed. Sal Restivo, 368–374. Oxford: Oxford University Press.

Ecks, Stefan. 2005. "Pharmaceutical Citizenship: Antidepressant Marketing and the Promise of Demarginalization in India." *Anthropology & Medicine*, 12, no. 3: 239–254. http://dx.doi.org/10.1080/13648470500291360.

Ecks, Stefan. 2008. "Three Propositions for an Evidence-Based Medical Anthropology." *Journal of the Royal Anthropological Institute*, 14, no. s1: S77–S92. http://dx.doi .org/10.1111/j.1467-9655.2008.00494.x.

Ecks, Stefan. 2010. "Polyspherical Pharmaceuticals: Global Psychiatry, Capitalism, and Space." In *Pharmaceutical Self: The Global Shaping of Experience in an Age of Psychopharmacology*, ed. Janis H. Jenkins, 97–116. Santa Fe, NM: School for Advanced Research Press.

Ehrenberg, Ronald G., Michael J. Rizzo, and Scott S. Condie. 2003. *Start-Up Costs in American Research Universities*. Cornell Higher Education Research Institute

Working Paper 33. Ithaca, NY: Cornell Higher Education Research Institute, Cornell University.

Elkana, Yehuda. 1981. "A Programmatic Attempt at an Anthropology of Knowledge." In *Sciences and Cultures: Anthropological and Historical Studies of the Sciences,* ed. Everett Mendelsohn and Yehuda Elkana, 1–71. Dordrecht, NL: Reidel.

Epstein, Gerald A., ed. 2005. *Financialization and the World Economy.* Cheltenham, UK: Edward Elgar.

Epstein, Steve. 2005. *Impure Science: AIDS, Activism, and the Politics of Knowledge.* Berkeley: University of California Press.

Eren Vural, Ipek. 2017. "Financialisation in Health Care: An Analysis of Private Equity Fund Investments in Turkey." *Social Science & Medicine,* 187 (Suppl. C): 276–286. https://doi.org/10.1016/j.socscimed.2017.06.008.

Ericson, Richard V., and Aaron Doyle, eds. 2003. *Risk and Morality.* Toronto: University of Toronto Press.

Etzkowitz, Henry. 2003. "Research Groups as 'Quasi-Firms': The Invention of the Entrepreneurial University." *Research Policy,* 32:109–121. http://dx.doi.org /10.1016/ S0048-7333(02)00009-4.

Etzkowitz, Henry. 2007. *MIT and the Rise of Entrepreneurial Science.* London: Routledge, Chapman & Hall.

Etzkowitz, Henry, Andrew Webster, Christiane Gebhardt, and Branca Regina Cantisano Terra. 2000. "The Future of the University and the University of the Future: Evolution of Ivory Tower to Entrepreneurial Paradigm." *Research Policy,* 29, no. 2: 313–330.

Etzkowitz, Henry, Andrew Webster, and Peter Healey, eds. 1998. *Capitalizing Knowledge: New Intersections of Industry and Academia.* Albany: SUNY Press.

Evans-Pritchard, E. E., and Eva Gillies. 1976. *Witchcraft, Oracles and Magic among the Azande.* New York: Oxford University Press.

Eyal, Nir, and Neema Sofaer. 2010. "The Diverse Ethics of Translational Research." *American Journal of Bioethics,* 10, no. 8: 19–30. http://dx.doi.org/10.1080 /15265161 .2010.494214.

Fabrega, Horacio. 1990. "An Ethnomedical Perspective of Medical Ethics." *Journal of Medicine & Philosophy,* 15, no. 6: 593–625. doi:10.1093/jmp/15.6.593.

Fang, Ferric C., and Arturo Casadevall. 2010. "Lost in Translation: Basic Science in the Era of Translational Research." *Infection and Immunity,* 78, no. 2: 563–566. http:// dx.doi.org/10.1128/IAI.01318-09.

Farquhar, Judith F., and John Kelly. 2013. "Commentary: Translation and Situation—Science, Metabolism, Finance." *Public Culture*, 25, no. 3 (71): 551–557. http://dx.doi.org/10.1215/08992363-214464.

Fassin, Didier. 2012. *Humanitarian Reason: A Moral History of the Present*, trans. Rachel Gomme. Berkeley: University of California Press.

Faulkner, Wendy. 1994. "Conceptualizing Knowledge Used in Innovation: A Second Look at the Science-Technology Distinction and Industrial Innovation." *Science, Technology, & Human Values*, 19, no. 4: 425–458. http://dx.doi.org/ 10.1177/ 016224399401900402.

Fins, Joseph J., Helen S. Mayberg, Bart Nuttin, Cynthia S. Kubu, Thorsten Galert, Volker Sturm, Katja Stoppenbrink, Reinhard Merkel, and Thomas E. Schlaepfer. 2011. "Misuse of the FDA's Humanitarian Device Exemption in Deep Brain Stimulation for Obsessive-Compulsive Disorder." *Health Affairs*, 30, no. 2: 302–311. http:// dx.doi.org/10.1377/hlthaff.2010.0157.

Fischer, Michael M. J. 2003. *Emergent Forms of Life and the Anthropological Voice*. Durham, NC: Duke University Press.

Fischer, Michael M. J. 2007. "Four Genealogies for a Recombinant Anthropology of Science and Technology." *Cultural Anthropology*, 22, no. 4: 539–615. http://dx.doi .org/10.1525/can.2007.22.4.539.

Fischer, Michael M. J. 2010. "Dr. Judah Folkman's Decalogue and Network Analysis." In *A Reader in Medical Anthropology: Theoretical Trajectories, Emergent Realities*, ed. Byron J. Good, Michael M. J. Fischer, Sarah S. Willen, and Mary-Jo DelVecchio Good, 339–344. Hoboken, NJ: Wiley-Blackwell.

Fischer, Michael M. J. 2012. "Lively Biotech and Translational Research." In *Lively Capital: Biotechnologies, Ethics, and Governance in Global Markets*, ed. Kaushik Sunder Rajan. Pp 385-436. Durham, NC: Duke University Press.

Fischer, Michael M. J. 2013. "Afterword: The Peopling of Technologies." In *When People Come First: Critical Studies in Global Health*, ed. João Biehl and Adriana Petryna, 347–374. Princeton, NJ: Princeton University Press.

Fischman, Josh. 2011. "New Psychiatric Surgeries Tread a Fine Line Between Research and Therapy." *Chronicle of Higher Education*, February 20. http://chronicle .com/article/New-Psychiatric-Surgeries/126463/.

Fitzgerald, Des. 2017. *Tracing Autism: Uncertainty, Ambiguity, and the Affective Labor of Neuroscience*. Seattle: University of Washington Press.

FitzGerald, Garret A. 2005. "Opinion: Anticipating Change in Drug Development: The Emerging Era of Translational Medicine and Therapeutics." *Nature Reviews Drug Discovery*, 4, no. 10: 815–818.

"5Minutes—*San Antonio Business Journal*." 2011. *Dennis A. Ahlburg Trinity University's president gives the 411 on The Center for the Sciences and Innovation*, April 22. http://www.bizjournals.com/sanantonio/print-edition/2011/04/22/5minutes.html ?page=all.

Fleck, Ludwick. 1981. *Genesis and Development of a Scientific Fact*. Chicago: University of Chicago Press.

Food and Drug Administration. n.d. "Humanitarian Device Exemption." *Medical Devices*. http://www.fda.gov/MedicalDevices/ DeviceRegulationandGuidance/ HowtoMarketYourDevice/PremarketSubmissions/HumanitarianDeviceExemption/.

Fortun, Michael. 2008. *Promising Genomics*. Berkeley: University of California Press.

Foucault, Michel. 1970. *The Order of Things: An Archaeology of the Human Sciences*. London: Tavistock.

Foucault, Michel. 1972. *Archaeology of Knowledge and the Discourse on Language*. London: Tavistock.

Fournier, Jay C., Robert J. DeRubeis, Steven D. Hollon, Sona Dimidjian, Jay D. Amsterdam, Richard C. Sheldon, and Jan Fawcett. 2010. "Antidepressant Drug Effects and Depression Severity: A Patient-Level Meta-Analysis." *Journal of the American Medical Association*, 303, no. 1: 47–53. http://dx.doi.org/ 10.1001/jama.2009.1943.

Franklin, Sarah. 2003. "Ethical Biocapital: New Strategies of Cell Culture." In *Remaking Life and Death: Toward an Anthropology of the Biosciences*, ed. Sarah Franklin and Margaret Lock, 97–128. Santa Fe, NM: School for Advanced Research Press.

Franklin, Sarah, and Margaret M. Lock. 2003. "Animation and Cessation: The Remaking of Life and Death." In *Remaking Life and Death: Toward an Anthropology of the Biosciences*, ed. Sarah Franklin and Margaret Lock, 1–30. Santa Fe, NM: School for Advanced Research Press.

Friedman, Richard A. 2013. "A Dry Pipeline for Psychiatric Drugs." *New York Times*, August 19. http://www.nytimes.com/2013/08/20/ health/a-dry-pipeline-for -psychiatric-drugs.html.

Friese, Carrie. 2013. "Realizing Potential in Translational Medicine: The Uncanny Emergence of Care as Science." *Current Anthropology*, 54, no. S7: S129–S138. https://doi.org/10.1086/670805.

Fujimura, Joan H. 1992. "Crafting Science: Standardized Packages, Boundary Objects, and 'translation.'" In *Science as Practice and Culture*, ed. Andrew Pickering, 168–211. Chicago: University of Chicago Press.

Fujimura, Joan H. 1996. *Crafting Science: A Sociohistory of the Quest for the Genetics of Cancer*. Cambridge, MA: Harvard University Press.

Fullwiley, Duana. 2011. *The Enculturated Gene: Sickle Cell Health Politics and Biological Difference in West Africa*. Princeton, NJ: Princeton University Press.

Gagnon, Marc-André, and Joel Lexchin. 2008. "The Cost of Pushing Pills: A New Estimate of Pharmaceutical Promotion Expenditures in the United States." *Public Library of Science Medicine*, 5, no. 1: e1. http://dx.doi.org/10.1371/journal.pmed.0050001.

Galison, Peter. 1997. *Image and Logic: A Material Culture of Microphysics*. Chicago: University of Chicago Press.

Galison, Peter. 1999. "Trading Zone: Coordinating Action and Belief." In *The Science Studies Reader*, ed. Mario Biagioli, 137–160. London: Routledge.

Galison, Peter, and Emily Thompson, eds. 1999. *The Architecture of Science*. Cambridge, MA: MIT Press.

Garcia-Rill, Edgar, ed. 2012. *Translational Neuroscience: A Guide to a Successful Program*. Chichester: Wiley.

Gardner, Howard. 1985. *The Mind's New Science: A History of the Cognitive Revolution*. New York: Basic Books.

Geertz, Clifford. 1978. "The Bazaar Economy: Information and Search in Peasant Marketing." *American Economic Review*, 68, no. 2: 28–32. http://www.jstor.org/stable/1816656.

Geertz, Clifford. 1983. *Local Knowledge: Further Essays in Interpretive Anthropology*. New York: Basic Books.

Geiger, R. L. 2002. "American Universities at the Beginning of the Twenty-First Century: Signposts on the Path to Privatization." In *Trends in American and German Higher Education*, ed. Robert Adams, 33–84. Cambridge, MA: American Academy of Arts and Sciences.

Gieryn, Thomas F. 1999. "Two Faces on Science: Building Identities for Molecular Biology and Biotechnology." In *The Architecture of Science*, ed. Peter Galison and Emily Thompson, 423–458. Cambridge, MA: MIT Press.

Gieryn, Thomas F. 2006. "City as Truth-Spot: Laboratories and Field-Sites in Urban Studies." *Social Studies of Science*, 36, no. 1: 5–38. http://dx.doi.org 10.1177/0306312705054526.

Gieryn, Thomas F. 2008. "Laboratory Design for Post-Fordist Science." *Isis*, 99, no. 4: 796–802. http://dx.doi.org/10.1086/595773.

Gilligan, Carol. 1993. *In a Different Voice: Psychological Theory and Women's Development*. Cambridge, MA: Harvard University Press.

Ginsberg, Steven. 2008. "Mission Bay Developers Take on Quake Challenges." *San Francisco Business Times*, October 26. http://www.bizjournals.com/sanfrancisco/stories/2008/10/27/focus6.html?page=all.

Gomez, Christopher M. 2006. "Translational Neuroscience: A Neurologist's Translation." *Current Neurology and Neuroscience Reports*, 6, no. 2: 85–87.

Good, Byron J. 1994. *Medicine, Rationality, and Experience.* Cambridge: Cambridge University Press.

Good, Byron J., Subandi, and Mary-Jo DelVecchio Good. 2007. "The Subject of Mental Illness: Psychosis, Mad Violence, and Subjectivity in Indonesia." In *Subjectivity: Ethnographic Investigations*, ed. João Biehl, Byron J. Good, and Arthur Kleinman, 243–272. Berkeley: University of California Press.

Good, Mary-Jo DelVecchio. 2007. "The Medical Imaginary and the Biotechnical Embrace: Subjective Experiences of Clinical Scientists and Patients." In *Subjectivity: Ethnographic Investigations*, ed. João Biehl, Byron J. Good, and Arthur Kleinman, 362–380. Berkeley: University of California Press.

Gordon, Deborah R. 1988. "Tenacious Assumptions in Western Medicine." In *Biomedicine Examined*, ed. Deborah R. Gordon and Margaret Lock, 19–57. Dordrecht, NL: Kluwer.

Grace, André P. 2012. "The Decline of Social Education and the Rise of Instrumentalism in North American Adult Education, 1947–1970." *Studies in the Education of Adults*, 44, no. 2: 225–244.

Graham, Janice E. 2006. "Diagnosing Dementia: Epidemiological and Clinical Data as Cultural Text." In *Thinking about Dementia: Culture, Loss, and the Anthropology of Senility*, ed. Annette Leibing and Lawrence Cohen, 80–105. New Brunswick, NJ: Rutgers University Press.

Greenslit, Nathan. 2002. "Pharmaceutical Branding: Identity, Individuality, and Illness." *Molecular Interventions*, 2, no. 6: 342–345. http://dx.doi.org/10.1124/mi.2.6.342.

Greenslit, Nathan P., and Ted J. Kaptchuk. 2012. "Antidepressants and Advertising: Psychopharmaceuticals in Crisis." *The Yale Journal of Biology and Medicine* 85 (1): 153–158.

Habermas, Jürgen. 1971. *Toward a Rational Society: Student Protest, Science and Politics.* London: Heinemann Educational.

Hackett, E. J. 2014. "Academic Capitalism." *Science, Technology, & Human Values*, 39, no. 5: 635–638. https://doi.org/10.1177/0162243914540219.

Hannaway, Owen. 1986. "Laboratory Design and the Aim of Science: Andreas Libavius versus Tycho Brahe." *Isis*, 77, no. 4: 585–610. doi:10.1086/354267.

Haraway, Donna J. 1989. *Primate Visions: Gender, Race, and Nature in the World of Modern Science*. New York: Routledge.

Haraway, Donna J., ed. 1990. "Situated Knowledges: The Science Question in Feminism and the Privilege of Partial Perspective." In *Simians, Cyborgs, and Women: The Reinvention of Nature, 575–599*. New York: Routledge.

Haraway, Donna J. 1996. "Modest Witness: Feminist Diffractions in Science Studies." In *The Disunity of Science: Boundaries, Contexts, and Power*, ed. Peter Galison and David J. Stump, 428–442. Stanford, CA: Stanford University Press.

Haraway, Donna J. 1997. *Modest_Witness@Second_Millennium.FemaleMan _Meets_ OncoMouse: Feminism and Technology*. New York: Routledge.

Harding, Sandra. 1993. "Rethinking Standpoint Epistemology: What Is 'Strong Objectivity'?" In *Feminist Epistemologies*, ed. Linda Alcoff and Elizabeth Potter, 49–82. New York: Routledge.

Harrington, Anne. 1989. *Medicine, Mind, and the Double Brain*. Princeton, NJ: Princeton University Press.

Harvey, David. 2005. *A Brief History of Neoliberalism*. Oxford: Oxford University Press.

Healy, David. 2004. "Psychopharmacology at the Interface between the Market and the New Biology." In *The New Brain Sciences: Perils and Prospects*, ed. Dai Rees and Steven Rose, 232–248. Cambridge: Cambridge University Press.

Healy, David. 2012. *Pharmageddon*. Berkeley: University of California Press.

Helmreich, Stefan. 2008. "Species of Biocapital." *Science as Culture*, 17, no. 4: 463–478. http://dx.doi.org/10.1080/09505430802519256.

Helmreich, Stefan, and Nicole Labruto. 2018. "Species of Biocapital, 2008, and Speciating Biocapital, 2017." In *The Palgrave Handbook of Biology and Society*, ed. Maurizio Meloni, John Cromby, Des Fitzgerald, and Stephanie Lloyd, 851–876. London: Palgrave Macmillan.

Herder, Matthew. 2013. "Emerging Academic Scientists' Exclusionary Encounters with Commercialization Law, Policy, and Practice." SSRN Scholarly Paper no. 2281609. Rochester, NY: Social Science Research Network. http://ssrn.com/abstract=2281609.

Herper, Matthew. 2012. "City Surgeon: Can the Cleveland Clinic Save Its Hometown?" *Forbes*, September 23. http://www.forbes.com /sites/matthewherper/2013/09/04/can-a-great-hospital-save-a-city.

Hess, David. 2001. "Ethnography and the Development of Science and Technology Studies." In *Handbook of Ethnography*, ed. Paul Atkinson, Amanda Coffey, Sara Delamont, John Lofland, and Lyn Lofland, 234–245. Thousand Oaks, CA: Sage.

Hinton, Ladson, Yvette Flores, Carol Franz, Isabel Hernandez, and Linda S. Mitteness. 2006. "The Borderlands of Primary Care: Physician and Family Perspectives on 'Troublesome' Behaviors of People with Dementia." In *Thinking about Dementia: Culture, Loss, and the Anthropology of Senility*, ed. Annette Leibing and Lawrence Cohen, 43–63. New Brunswick, NJ: Rutgers University Press.

Ho, Karen. 2009. *Liquidated: An Ethnography of Wall Street*. Durham, NC: Duke University Press.

Ho, Karen. 2018. "Markets, Myths, and Misrecognitions: Economic Populism in the Age of Financialization and Hyperinequality." *Economic Anthropology*, 5, no. 1: 148–150. https://doi.org/10.1002/sea2.12112.

Hofstadter, Richard. 1963. *Anti-Intellectualism in American Life*. New York: Knopf.

Holman, Bennett, and Kevin C. Elliott. 2018. "The Promise and Perils of Industry-Funded Science." *Philosophy Compass* 0 (0): e12544. https://doi.org/10.1111/phc3.12544.

Hong, Wei, and John P. Walsh. 2009. "For Money or Glory? Commercialization, Competition, and Secrecy in the Entrepreneurial University." *Sociological Quarterly*, 50, no. 1: 145–171. http://dx.doi.org/10.1111/j.1533-8525.2008.01136.x.

Hopkins, Michael M., Philippa A. Crane, Paul Nightingale, and Charles Baden-Fuller. 2013. "Buying Big into Biotech: Scale, Financing, and the Industrial Dynamics of UK Biotech, 1980–2009." *Industrial and Corporate Change*, 22 (August): 903–952.

Horvath, Michael. 2004. "Regional Comparisons of Innovation Activity Based on Venture Capital Flows." In *Building High-Tech Clusters: Silicon Valley and Beyond*, ed. Timothy Bresnahan and Alfonso Gambardella, 280–330. Cambridge: Cambridge University Press.

Hughes, Sally Smith. 2011. *Genentech: The Beginnings of Biotech*. Chicago: University of Chicago Press.

Insel, Thomas R. 2009. "Disruptive Insights in Psychiatry: Transforming a Clinical Discipline." *Journal of Clinical Investigation*, 119, no. 4: 700–705. http://dx.doi.org/10.1172/JCI38832.

Insel, Thomas R. 2010. Quoted in Greg Miller, "Is Pharma Running Out of Brainy Ideas? Recent Cutbacks Raise Concerns about the Future of Drug Development for Nervous System Disorders." *Science*, 329: 502–504. http://dx.doi.org/ 10.1126/science.329.5991.502.

Ioannidis, John P. A. 2004. "Materializing Research Promises: Opportunities, Priorities and Conflicts in Translational Medicine." *Journal of Translational Medicine*, 2, no. 5: 1–6. http://dx.doi.org/10.1186/1479-5876-2-5.

Jacobides, Michael G. 2005. "Industry Change through Vertical Disintegration: How and Why Markets Emerged in Mortgage Banking." *Academy of Management Journal*, 48, no. 3: 465–498.

Jain, Sarah S. Lochlann. 2006. Injury: The Politics of Product Design and Safety Law in the United States. Princeton, NJ: Princeton University Press.

Jain, Sarah S. Lochlann. 2013. Malignant: How Cancer Becomes Us. Berkeley: University of California Press.

Jameson, Fredric, and Masao Miyoshi. 1998. The Cultures of Globalization. Durham, NC: Duke University Press.

Jarvis, Lisa M. 2012. "Pfizer's Academic Experiment." Chemical & Engineering News, 90, no. 40: 28–32.

Jasanoff, Sheila, and Sang-Hyun Kim. 2013. "Sociotechnical Imaginaries and National Energy Policies." Science as Culture, 22, no. 2: 189–196. http://dx.doi.org/10.1080/09505431.2013.786990.

Jenkins, Janis H. 2011. "Pharmaceutical Self and Imaginary in the Social Field of Psychiatric Treatment." In Pharmaceutical Self: The Global Shaping of Experience in an Age of Psychopharmacology, ed. Janis H. Jenkins, 17–40. Santa Fe, NM: School for Advanced Research Press.

Jogalekar, Ashutosh. 2012. "The Perils of Translational Research." The Curious Wavefunction, Scientific American Blog Network, November 26. *Scientific American*. http://blogs.scientificamerican.com/the-curious-wavefunction/2012/11/26/the-perils-of-translational-research/.

Johnson & Johnson Innovation Centers. n.d. https://www.jnj.com/innovation-at-jnj.

Joyce, Kelly. 2005. "Appealing Images: Magnetic Resonance Imaging and the Production of Authoritative Knowledge." *Social Studies of Science*, 35, no. 3: 437–462. http://dx.doi.org/10.1177/0306312705050180.

Joyce, Kelly. 2008. *Magnetic Appeal: MRI and the Myth of Transparency*. Ithaca, NY: Cornell University Press.

Kahn, Jonathan. 2012. *Race in a Bottle: The Story of BiDil and Racialized Medicine in a Post-Genomic Age*. New York: Columbia University Press.

Kamat, Vinay R., and Mark Nichter. 1998. "Pharmacies, Self-Medication and Pharmaceutical Marketing in Bombay, India." *Social Science & Medicine*, 47, no. 6: 779–794. doi:10.1016/S0277-9536(98)00134-8.

Kaplan, Aaron V., Elisa D. Harvey, Richard E. Kuntz, Hadas Shiran, John F. Robb, and Peter Fitzgerald. 2005. "Humanitarian Use Devices/Humanitarian Device

Exemptions in Cardiovascular Medicine." *Circulation*, 112, no. 18: 2883–2886. http://dx.doi.org/10.1161/CIRCULATIONAHA.105.553701.

Karlin, Jennifer. 2013. "Loss and Gain in Translation: Financial Epidemiology on the South Side of Chicago." *Public Culture*, 25, no. 3: 523–550. http://dx.doi.org/10.1215/08992363-2144634.

Kaufman, Sharon R. 2006. "Dementia-Near-Death and 'Life Itself.'" In *Thinking about Dementia: Culture, Loss, and the Anthropology of Senility*, ed. Annette Leibing and Lawrence Cohen, 23–42. New Brunswick, NJ: Rutgers University Press.

Keating, Peter K., and Alberto Cambrosio. 2000. "'Real Compared to What?' Diagnosing Leukemias and Lymphomas." In *Living and Working with New Medical Technologies*, ed. Margaret M. Lock, Allan Young, and Alberto Cambrosio, 103–134. Cambridge: Cambridge University Press.

Keating, Peter K., and Alberto Cambrosio. 2003. *Biomedical Platforms: Realigning the Normal and the Pathological in Late-Twentieth-Century Medicine*. Cambridge, MA: MIT Press.

Keller, Evelyn Fox. 1995. *Reflections on Gender and Science*. New Haven, CT: Yale University Press.

Kelty, Christopher M., Michael M. J. Fischer, Alex "Rex" Golub, Jason Baird Jackson, Kimberly Christen, Michael F. Brown, and Tom Boellstorff. 2008. "Anthropology of/ in Circulation: The Future of Open Access and Scholarly Societies." *Cultural Anthropology*, 23, no. 3: 559–588. http://dx.doi.org/10.1525/can.2008.23.3.559.

Kenney, Martin. 1986. *Biotechnology: The University-Industrial Complex*. Yale University Press.

Kenney, Martin, and Richard Florida. 2000. "Venture Capital in Silicon Valley: Fueling New Firm Creation." In *Understanding Silicon Valley: The Anatomy of an Entrepreneurial Region*, ed. Martin Kenney, 98–123. Stanford, CA: Stanford University Press.

Kesselheim, Aaron S., Michelle M. Mello, and David M. Studdert. 2011. "Strategies and Practices in Off-Label Marketing of Pharmaceuticals: A Retrospective Analysis of Whistleblower Complaints." *PLOS Medicine*, 8, no. 4: e1000431. http://dx.doi.org/10.1371/journal.pmed.1000431.

Kessler, Suzanne J. 1998. *Lessons from the Intersexed*. New Brunswick, NJ: Rutgers University Press.

Kingery, W. David. 2001. "The Design Process as a Critical Component of the Anthropology of Technology." In *Anthropological Perspectives on Technology*, ed. Michael B. Schiffer pp. 123-138. Albuquerque: University of New Mexico Press.

Kingfisher, Catherine, and Jeff Maskovsky. 2008. "Introduction: The Limits of Neoliberalism." *Critique of Anthropology*, 28, no. 2: 115–126. http://dx.doi.org/10.1177/0308275X08090544.

Kirmayer, Laurence J. 2012. "The Future of Critical Neuroscience." In *Critical Neuroscience: A Handbook of the Social and Cultural Contexts of Neuroscience,* ed. Suparna Choudhury and Jan Slaby, 367–383. Chichester: Wiley-Blackwell.

Kirmayer, Laurence J., and Eugene Raikhel. 2009. "From Amrita to Substance D: Psychopharmacology, Political Economy, and Technologies of the Self" (editorial). *Transcultural Psychiatry*, 46, no. 1: 5–15. http://dx.doi.org/10.1177/1363461509102284.

Kirsch, Irving, Brett J. Deacon, Tania B. Huedo-Medina, Alan Scoboria, Thomas J. Moore, and Blair T. Johnson. 2008. "Initial Severity and Antidepressant Benefits: A Meta-Analysis of Data Submitted to the Food and Drug Administration." *PLOS Medicine* 5 (2): e45. https://doi.org/10.1371/journal.pmed.0050045.

Kirsh, David, and Paul Maglio. 1994. "On Distinguishing Epistemic from Pragmatic Action." *Cognitive Science*, 18, no. 4: 513–549. http://dx.doi.org/10.1207/s15516709cog1804_1.

Klein, Julie Thompson. 1996. *Crossing Boundaries: Knowledge, Disciplinarities, and Interdisciplinarities*. Charlottesville: University of Virginia Press.

Kleinman, Daniel Lee. 2003. *Impure Cultures: University Biology and the World of Commerce*. Madison: University of Wisconsin Press.

Kleinman, Daniel Lee, and Steven P. Vallas. 2001. "Science, Capitalism, and the Rise of the 'Knowledge Worker': The Changing Structure of Knowledge Production in the United States." *Theory and Society*, 30, no. 4: 451–492.

Knorr-Cetina, Karin D. 1981. *The Manufacture of Knowledge: An Essay on the Constructivist and Contextual Nature of Science*. Oxford: Pergamon Press.

Knorr-Cetina, Karin D. 1992. "The Couch, the Cathedral, and the Laboratory: On the Relationship between Experiment and Laboratory in Science." In *Science as Practice and Culture*, ed. Andrew Pickering, 113–138. Chicago: University of Chicago Press.

Knorr-Cetina, Karin D. 1999. *Epistemic Cultures: How the Sciences Make Knowledge*. Cambridge, MA: Harvard University Press.

Knowledge@Wharton.com. 2009. "Re-thinking Risk Management: Why the Mindset Matters More Than the Model." *Knowledge@Wharton, University of Pennsylvania*, April 15. http://knowledge.wharton.upenn.edu/article/re-thinking-risk-management-why-the-mindset-matters-more-than-the-model/.

Kohler, Robert E. 2002. *Landscapes and Labscapes: Exploring the Lab-Field Border in Biology*. Chicago: University of Chicago Press.

Krimsky, Sheldon. 2004. *Science in the Private Interest: Has the Lure of Profits Corrupted Biomedical Research?* Lanham, MD: Rowman & Littlefield.

Kringelbach, Morton L., Ned Jenkinson, Sarah L. F. Owen, and Tipu Z. Aziz. 2007. "Translational Principles of Deep Brain Stimulation." *Nature Reviews Neuroscience*, 8, no. 8: 623–635.

Kuhn, Thomas S. 1996. *The Structure of Scientific Revolutions*. 3rd ed. Chicago: University of Chicago Press. (Orig. pub. 1962.)

Kurtzman, Laura. 2018. "UCSF, Pfizer Renew Research Collaboration, Citing Progress in Drug Discovery Research." *UC San Francisco News Center*, January 9. https://www.ucsf.edu/news/2017/01/405461/ucsf-pfizer-renew-research-collaboration-citing-progress-drug-discovery-research.

Lakoff, Andrew. 2005. *Pharmaceutical Reason: Knowledge and Value in Global Psychiatry*. New York: Cambridge University Press.

Lam, Alice. 2010. "From 'Ivory Tower Traditionalists' to 'Entrepreneurial Scientists'? Academic Scientists in Fuzzy University-Industry Boundaries." *Social Studies of Science*, 40, no. 2: 307–340. http://dx.doi.org/10.1177/0306312709349963.

Lamont, Michèle. 2009. *How Professors Think: Inside the Curious World of Academic Judgment*. Cambridge, MA: Harvard University Press.

Langley, Paul. 2018. "Frontier Financialization: Urban Infrastructure in the United Kingdom." *Economic Anthropology*, 5, no. 2: 172–184. https://doi.org/10.1002/sea2.12115.

Langlitz, Nicholas. 2010. "Biosciences, Ethics and Collaborations." *BioSocieties*, 5, no. 3: 391–392. http://dx.doi.org/10.1057/biosoc.2010.20.

Latour, Bruno. 1983. "Give Me a Laboratory and I Will Raise the World." In *Science Observed*, ed. Karin Knorr-Cetina and Michael Joseph Mulkay, 141–170. London: Sage.

Latour, Bruno. 1988. *Science in Action: How to Follow Scientists and Engineers through Society*. Cambridge, MA: Harvard University Press.

Latour, Bruno. 1991. "Technology Is Society Made Durable" (Sociological Review Monograph no. 38). In *A Sociology of Monsters: Essays on Power, Technology and Domination*, ed. John Law, 103–132. http://www.bruno-latour.fr/sites/default/files/46-TECHNOLOGY-DURABLE-GBpdf.pdf.

Latour, Bruno. 2005. *Reassembling the Social: An Introduction to Actor-Network Theory*. Oxford: Oxford University Press.

Latour, Bruno. 2008. "A 'Cautious Prometheus'? A Few Steps toward a Philosophy of Design." Keynote address, Networks of Design International Conference, Falmouth University, September 3.

Latour, Bruno, and Steve Woolgar. 1986. *Laboratory Life*. Princeton, NJ: Princeton University Press.

Lauto, Giancarlo, and Finn Valentin. 2016. "The Knowledge Production Model of the New Sciences: The Case of Translational Medicine." *Technological Forecasting and Social Change* 111 (C): 12–21.

Lave, Rebecca, Philip Mirowski, and Samuel Randalls. 2010. "Introduction: STS and Neoliberal Science." *Social Studies of Science*, 40, no. 5: 659–675. http://dx.doi.org/10.1177/0306312710378549.

Law, John. 2009. "Actor Network Theory and Material Semiotics." In *The New Blackwell Companion to Social Theory*, ed. Bryan S. Turner, 143–157. Chichester: Wiley.

Lee, Peter. 2012. "Transcending the Tacit Dimension: Patents, Relationships, and Organizational Integration in Technology Transfer." SSRN Scholarly Paper no. ID 2019335. *Social Science Research Network*. http://papers.ssrn.com/abstract =2019335.

Lee, Thomas. 2010. "FDA Calls Out Medtronic on DBS Device." *MassDevice–Medical Device Industry News*, March 11. http://www.massdevice.com/news/fda-calls-out-medtronic-dbs-device.

Lee, Tih-Shih, Beng-Yeong Ng, and Wei-Ling Lee. 2008. "Neuropsychiatry: An Emerging Field." *Annals of the Academy of Medicine, Singapore*, 37, no. 7: 601–605.

Lefebvre, Henri. 1991. *The Production of Space*. Oxford: Wiley.

Leibing, Annette. 2009. "Tense Prescriptions? Alzheimer Medications and the Anthropology of Uncertainty." *Transcultural Psychiatry*, 46, no. 1: 180–206. http://dx.doi.org/10.1177/1363461509102297.

Le Merle, Matthew, and Jamie Campbell. 2011. "Building an External Innovation Capability." Booz & Allen white paper. http://www.strategyand.pwc.com /global/home/what-we-think/reports-white-papers/article-display/building-external-innovation-capability.

Light, Donald W., and Joel R. Lexchin. 2012. "Pharmaceutical Research and Development: What Do We Get for All That Money?" *BMJ* (Clinical Research Ed.), 345:e4348.

Light, Donald W., Joel R. Lexchin, and Jonathan J. Darrow. 2013. "Institutional Corruption of Pharmaceuticals and the Myth of Safe and Effective Drugs." SSRN Scholarly Paper no. 2282014. *Social Science Research Network*. http://papers.ssrn.com/abstract=2282014.

Light, Donald W., and Antonio F. Maturo. 2015. *Good Pharma: The Public-Health Model of the Mario Negri Institute.* New York: Palgrave Macmillan.

Lippman, Abby. 1992. "Led (Astray) by Genetic Maps: The Cartography of the Human Genome and Human Care." *Social Science & Medicine*, 35, no. 12: 1469–1476.

Lock, Margaret M. 2002. *Twice Dead.* Berkeley: University of California Press.

Lock, Margaret M. 2007. "The Future Is Now: Locating Biomarkers for Dementia." In *Biomedicine as Culture: Instrumental Practices, Technoscientific Knowledge, and New Modes of Life*, ed. Regula Valérie Burri and Joseph Dumit, 6–61. New York: Routledge.

Lock, Margaret M. 2013. *The Alzheimer Conundrum: Entanglements of Dementia and Aging.* Princeton, NJ: Princeton University Press.

Lock, Margaret M., and Deborah Gordon. 1988. *Biomedicine Examined.* Dordrecht, NL: Kluwer Academic.

Lock, Margaret M., and Vinh-Kim Nguyen. 2010. *An Anthropology of Biomedicine.* Chichester: Wiley-Blackwell.

Lock, Margaret M., and Mark Nichter. 2002. "Introduction: From Documenting Medical Pluralism to Critical Interpretations of Globalized Health Knowledge, Policies, and Practices." In *New Horizons in Medical Anthropology: Essays in Honour of Charles Leslie*, ed. Margaret M. Lock and Mark Nichter, 1–34. London: Routledge.

Longino, Helen E. 1990. *Science as Social Knowledge: Values and Objectivity in Scientific Inquiry.* Princeton, NJ: Princeton University Press.

Lovell, Anne M. 2007. "Hoarders and Scrappers: Madness and the Social Person in the Interstices of the City." In *Subjectivity: Ethnographic Investigations*, ed. João Biehl, Byron J. Good, and Arthur Kleinman, 315–340. Berkeley: University of California Press.

Lowen, Rebeca S. 1997. *Creating the Cold War University: The Transformation of Stanford.* Berkeley: University of California Press.

Löwy, Ilana. 2000. "Trustworthy Knowledge and Desperate Patients: Clinical Tests for New Drugs from Cancer to AIDS." In *Living and Working with New Medical Technologies*, ed. Margaret Lock, Allan Young, and Alberto Cambrosio, 49–81. Cambridge: Cambridge University Press.

Löwy, Ilana, and Emilia Sanabria. 2016. "The Biomedicalization of Brazilian Bodies: Anthropological Perspectives." *História, Ciências, Saúde-Manguinhos*, 23, no. 1: 11–18.

Luhmann, Niklas. 1993. "Deconstruction as Second-Order Observing." *New Literary History*, 24, no. 4: 763–782. http://www.jstor.org/stable/469391.

Lynch, Zack, and Byron Laursen. 2010. *The Neuro Revolution: How Brain Science Is Changing Our World.* New York: St. Martin's Griffin.

Lyotard, Jean-François. 1984. *The Postmodern Condition: A Report on Knowledge*. Trans. Geoff Bennington and Brian Massuni. Minneapolis: University of Minnesota Press.

MacDonald, Ann H. 2011. "Deep Brain Stimulation: Experts Raise Alarms about Aggressive Marketing." *Harvard Health Blog*, February 15. http://www.health.harvard.edu/blog/deep-brain-stimulation-experts-raise-alarms-about-aggressive-marketing-201102151480.

Mack, George S. 2013. "A Compulsion for Brain Science: Zack Lynch." *Streetwise Reports*, October 17. http://www.thelifesciencesreport.com/pub/na/15666.

Maienschein, Jane, Mary Sunderland, Rachel A. Ankeny, and Jason Scott Robert. 2008. "The Ethos and Ethics of Translational Research." *American Journal of Bioethics: AJOB*, 8, no. 3: 43–51. https://doi.org/10.1080/15265160802109314.

Marincola, Francesco M. 2003. "Translational Medicine: A Two-Way Road." *Journal of Translational Medicine*, 1, no. 1: 1. http://dx.doi.org/10.1186/1479-5876-1-1.

Martin, Emily. 1991. "The Egg and the Sperm: How Science Has Constructed a Romance Based on Stereotypical Male-Female Roles." *Signs: Journal of Women in Culture and Society*, 16, no. 3: 485–501.

Martin, Emily. 2007. *Bipolar Expeditions: Mania and Depression in American Culture*. Princeton, NJ: Princeton University Press.

Matheson, Alastair. 2008. "Corporate Science and the Husbandry of Scientific and Medical Knowledge by the Pharmaceutical Industry." *BioSocieties*, 3, no. 4: 355–382. http://dx.doi.org/10.1017/S1745855208006297.

McGoey, Linsey. 2016. *No Such Thing as a Free Gift: The Gates Foundation and the Price of Philanthropy*. London: Verso.

Merz, Martina, and Karin Knorr-Cetina. 1997. "Deconstruction in a 'Thinking' Science: Theoretical Physicists at Work." *Social Studies of Science*, 27, no. 1: 73–111. http://pub.uni-bielefeld.de/publication/2399834.

Miller, Greg. 2010. "Is Pharma Running Out of Brainy Ideas? Recent Cutbacks Raise Concerns about the Future of Drug Development for Nervous System Disorders." *Science*, 329, no. 5991: 502–504. http://dx.doi.org/10.1126/science.329.5991.502.

Minogue, Kristen, and Howard Wolinsky. 2010. "Lost in Translation." *European Molecular Biology Organization Reports*, 11, no. 2: 93–96. http://dx.doi.org/ 10.1038/ embor.2009.282.

Mirowski, Philip. 2011. *Science-Mart: Privatizing American Science*. Cambridge, MA: Harvard University Press.

Mirowski, Philip. 2012. "The Modern Commercialization of Science Is a Passel of Ponzi Schemes." *Social Epistemology*, 26, no. 3–4: 285–310. https://doi.org/10.1080/ 02691728.2012.697210.

Mittra, James. 2015. *The New Health Bioeconomy: R&D Policy and Innovation for the Twenty-First Century*. New York: Palgrave Macmillan.

Miyazaki, Hirokazu, and Annelise Riles. 2005. "Failure as Endpoint." In *Global Assemblages: Technology, Politics, and Ethics as Anthropological Problems*, ed. Aihwa Ong and Stephen J. Collier, 320–331. Malden, MA: Blackwell.

Moerman, Daniel E. 2002. *Meaning, Medicine, and the "Placebo Effect."* Cambridge: Cambridge University Press.

Mol, Annemarie. 2002. *The Body Multiple: Ontology in Medical Practice*. Durham, NC: Duke University Press.

Molé, Noelle J. 2012. *Labor Disorders in Neoliberal Italy: Mobbing, Well-Being, and the Workplace*. Bloomington: Indiana University Press.

Moncrieff, Joanna. 2006. "Psychiatric Drug Promotion and the Politics of Neo-Liberalism." *British Journal of Psychiatry*, 188: 301–302. http://dx.doi.org /10.1192/ bjp.188.4.301.

Montgomery, Erwin B. Jr., and John T. Gale. 2008. "Mechanisms of Action of Deep Brain Stimulation (DBS)." *Neuroscience & Biobehavioral Reviews*, 32, no. 3: 388–407.

Mowery, David C., Richard R. Nelson, Bhaven N. Sampat, and Arvidsa Ziedonis. 2004. *Ivory Tower and Industrial Innovation: University-Industry Technology Transfer Before and After the Bayh-Dole Act*. Stanford, CA: Stanford University Press.

National Center for Advancing Translational Sciences. 2013. "NIH to Fund Collaborations with Industry to Identify New Uses for Existing Compounds." US Department of Health and Human Services, June 18. http://www.nih.gov/news/health/ jun2013/ncats-18.htm.

National Institute of Mental Health. n.d. "The Numbers Count: Mental Disorders in America." http://www.nimh.nih.gov/health /publications/the-numbers-count -mental-disorders-in-america/index.shtml.

Nelson, Alondra. 2013. *Body and Soul: The Black Panther Party and the Fight against Medical Discrimination*, 1st ed. Minneapolis: University of Minnesota Press.

Nelson, Alondra. 2016. *The Social Life of DNA: Race, Reparations, and Reconciliation After the Genome*. Boston: Beacon Press.

NeuroInsights. 2009. *The Neurotechnology Industry 2009 Report*. San Francisco: NeuroInsights.

NeuroInsights. 2010. *The Neurotechnology Industry 2010 Report*. San Francisco: NeuroInsights.

Nichter, Mark, and Nancy Vuckovic. 1994. "Agenda for an Anthropology of Pharmaceutical Practice." *Social Science & Medicine*, 39, no. 11: 1509–1525.

Ninnemann, Kristi M. 2012. "Variability in the Efficacy of Psychopharmaceuticals: Contributions from Pharmacogenomics, Ethnopsychopharmacology, and Psychological and Psychiatric Anthropologies." *Culture, Medicine, and Psychiatry*, 36, no. 1: 10–25. http://dx.doi.org/10.1007/s11013-011-9242-y.

Nutt, David, and Guy Goodwin. 2011. "ECNP Summit on the Future of CNS Drug Research in Europe 2011." *European Neuropsychopharmacology*, 21:495–499. http://dx.doi.org/10.1016/j.euroneuro.2011.05.004.

Nye, David E. 2007. *Technology Matters: Questions to Live With*. Cambridge, MA: MIT Press.

Ofstehage, Andrew L. 2018. "Financialization of Work, Value, and Social Organization among Transnational Soy Farmers in the Brazilian Cerrado." *Economic Anthropology*, 5, no. 2: 274–285. https://doi.org/10.1002/sea2.12123.

Olssen, Mark, and Michael A. Peters. 2005. "Neoliberalism, Higher Education and the Knowledge Economy: From the Free Market to Knowledge Capitalism." *Journal of Education Policy*, 20, no. 3: 313–345. http://dx.doi.org/10.1080/ 02680930500108718.

Raaper, Rille, and Mark Olssen. 2015. "Mark Olssen on Neoliberalisation of Higher Education and Academic Lives: An Interview." *Policy Futures in Education* 14 (2): 147–163. https://doi.org/10.1177/1478210315610992.

Ong, Aihwa. 2010. "Introduction." In *Asian Biotech: Ethics and Communities of Fate*, ed. Aihwa Ong and Nancy N. Chen, 1–44. Durham, NC: Duke University Press.

Ong, Aihwa, and Stephen J. Collier. 2005. "Global Assemblages, Anthropological Problems." In *Global Assemblages*, ed. Aihwa Ong and Stephen J. Collier, 3–21. Malden, MA: Blackwell.

Orr, Liesl. 1997. "Globalisation and Universities: Towards the 'Market University'?" *Social Dynamics: A Journal of African Studies*, 23, no. 1: 42–67. http://dx.doi. org/10.1080 /02533959708458619.

Peña, Carlos, Kristen Bowsher, Ann Costello, Robert De Luca, Sara Doll, Khan Li, Marie Schroeder, and Theodore Stevens. 2007. "An Overview of FDA Medical Device Regulation as It Relates to Deep Brain Stimulation Devices." *IEEE Transactions on Neural Systems and Rehabilitation Engineering: A Publication of the IEEE Engineering in Medicine and Biology Society*, 15, no. 3: 421–424.

Petryna, Adriana. 2005. "Ethical Variability: Drug Development and Globalizing Clinical Trials." *American Ethnologist*, 32, no. 2: 183–197. http://dx.doi.org/10.1525 / ae.2005.32.2.183.

Petryna, Adriana. 2006. "Globalizing Human Subjects Research." In *Global Pharmaceuticals: Ethics, Markets, Practices*, ed. Adriana Petryna, Andrew Lakoff, and Arthur Kleinman, 33–60. Durham, NC: Duke University Press.

Petryna, Adriana. 2009. *When Experiments Travel: Clinical Trials and the Global Search for Human Subjects.* Princeton, NJ: Princeton University Press.

Petryna, Adriana, and Arthur Kleinman. 2006. "The Pharmaceutical Nexus." In *Global Pharmaceuticals: Ethics, Markets, Practices,* ed. Adriana Petryna, Andrew Lakoff, and Arthur Kleinman, 1–32. Durham, NC: Duke University Press.

Petryna, Adriana, Andrew Lakoff, and Arthur Kleinman, eds. 2006. *Global Pharmaceuticals: Ethics, Markets, Practices.* Durham, NC: Duke University Press.

Pfaffenberger, Bryan. 1992. "Technological Dramas." *Science, Technology & Human Values,* 17, no. 3: 282–312. http://dx.doi.org/10.1177/016224399201700302.

Pfaffenberger, Bryan. 2001. "Symbols Do Not Create Meanings—Activities Do: Or, Why Symbolic Anthropology Needs the Anthropology of Technology." In *Anthropological Perspectives on Technology,* ed. Michael Brian Schiffer, 77–86. Albuquerque: University of New Mexico Press.

Pfotenhauer, Sebastian, and Sheila Jasanoff. 2017. "Traveling Imaginaries: The 'Practice Turn' in Innovation Policy and the Global Circulation of Innovation Models." In *The Routledge Handbook of the Political Economy of Science,* ed. David Tyfield, Rebecca Lave, Samuel Randalls, and Charles Thorpe. Oxford: Routledge, 416–428

Pickersgill, Martyn D. 2013. "From 'Implications' to 'Dimensions': Science, Medicine and Ethics in Society." *Health Care Analysis,* 21, no. 1: 31–42. http://dx.doi.org/10.1007/s10728-012-0219-y.

Pinch, Trevor, and Wiebe E. Bijker. 1984. "The Social Construction of Facts and Artifacts: Or How the Sociology of Science and the Sociology of Technology Might Benefit Each Other." *Social Studies of Science,* 14:399–441. http://www.alice.id.tue.nl/references/pinch-bijker-1984.pdf.

Pisano, Gary. P. 2006. *Science Business: The Promise, the Reality, and the Future of Biotech.* Boston: Harvard Business School Press.

Pitluck, Aaron Z., Fabio Mattioli, and Daniel Souleles. 2018. "Finance beyond Function: Three Causal Explanations for Financialization." *Economic Anthropology,* 5, no. 2: 157–171. https://doi.org/10.1002/sea2.12114.

Polanyi, Karl. 2001. *The Great Transformation: The Political and Economic Origins of Our Time.* Boston: Beacon Press. (Orig. pub. 1944.)

Pollock, Anne. 2008. "Pharmaceutical Meaning-Making beyond Marketing: Racialized Subjects of Generic Thiazide." *Journal of Law, Medicine & Ethics,* 36, no. 3: 530–536. doi:10.1111/j.1748-720X.2008.301.x.

Poon, Martha A. 2008. "From New Deal Institutions to Capital Markets: Commercial Consumer Risk Scores and the Making of Subprime Mortgage Finance." SSRN

Scholarly Paper no. 1458545. *Social Science Research Network*. http://papers.ssrn.com/abstract=1458545.

Poon, Martha A. 2012. "Why Does Finance Need an Anthropology? Because Financial Value Is a Reality." *Cultural Anthropology*, May 12. http://www.culanth.org/fieldsights/358-why-does-finance-need-an-anthropology-because-financial-value-is-a-reality.

Power, Michael. 1999. *The Audit Society: Rituals of Verification*. Oxford: Oxford University Press.

Pritchard, Justine R., Peter M. Bruno, Luke A. Gilbert, Kelsey Capron, Douglas A. Lauffenburger, and Michael T. Hemann. 2013. "Defining Principles of Combination Drug Mechanisms of Action." *Proceedings of the National Academy of Sciences of the United States of America*, 110, no. 2: e170–e179. http://dx.doi.org/ 10.1073/pnas.1210419110.

Raaper, Rille, and Mark Olssen. 2015. "Mark Olssen on Neoliberalisation of Higher Education and Academic Lives: An Interview." *Policy Futures in Education*, 14, no. 2: 147–163. https://doi.org/10.1177/1478210315610992.

Rabinow, Paul. 1997. *Making PCR: A Story of Biotechnology*. Chicago: University of Chicago Press.

Rabinow, Paul. 2002. *French DNA: Trouble in Purgatory*. Chicago: University of Chicago Press.

Rabinow, Paul, and Gaymon Bennett. 2012. *Designing Human Practices: An Experiment with Synthetic Biology*. Chicago: University of Chicago Press.

Racine, Eric. 2010. *Pragmatic Neuroethics: Improving Treatment and Understanding of the Mind-Brain*. Cambridge, MA: MIT Press.

Rapp, Rayna. 2011. "A Child Surrounds This Brain: The Future of Neurological Difference According to Scientists, Parents, and Diagnosed Young Adults." In *Sociological Reflections on the Neurosciences*, ed. Martyn Pickersgill and Ira Van Keulen, 3–26. Bingley, UK: Emerald Group Publishing.

Readings, Bill. 1997. *The University in Ruins*. Cambridge, MA: Harvard University Press.

Reed, John C., E. Lucile White, Jeffrey Aubé, Craig Lindsley, Min Li, Larry Sklar, and Stuart Schreiber. 2012. "The NIH's Role in Accelerating Translational Sciences." *Nature Biotechnology*, 30, no. 1: 16–19. http://dx.doi.org/10.1038/nbt.2087.

Rees, Dai, and Steven Rose. 2004. *The New Brain Sciences: Perils and Prospects*. Cambridge: Cambridge University Press.

Reisch, Marc. 2011. "Novartis to Cut 2,000 Jobs." *Chemical & Engineering News*, 89 (October 28, 2011), no. 44. https://pubs.acs.org/cen/news/89/i44/8944notw9.html.

Restivo, Sal P. 2005. *Science, Technology, and Society*. New York: Oxford University Press.

Riles, Annelise. 2006. *Documents: Artifacts of Modern Knowledge*. Ann Arbor: University of Michigan Press.

Riskviews. 2011. "The Difference between Risk & Loss." *Riskviews* (blog). March 3, 2011. https://riskviews.wordpress.com/2011/03/02/the-difference-between-risk-loss.

Robinson, Mark. 2017. "Translational Medicine: Science, Risk and an Emergent Political Economy of Biomedical Innovation." In *The Routledge Handbook of the Political Economy of Science*, ed. David Tyfield, Rebecca Lave, Samuel Randalls, and Charles Thorpe, 249–260. New York: Routledge.

Robinson, Mark D. 2018. "Financializing Epistemic Norms in Contemporary Biomedical Innovation." *Synthese*, February. https://doi.org/10.1007/s11229-018-1704-0.

Robinson, Mark. (in preparation a). "Engineering the Translational Imagination: Instrumentalism and Ideation in Translational Research."

Robinson, Mark. (in preparation b) "A Modern History of Pharmaceutical Innovation and the Birth of Translational Medicine." in *Medicines, Histories and Translations*. Eds. Carsten Timmerman, Michael Worboys, and Elizabeth Toon. London: Palgrave Macmillan

Rodríguez, José. 2011. "Campus Dedicates Li Ka Shing Center for Biomedical and Health Sciences, Philanthropist Receives Berkeley Medal." *News, Berkeley Research*, October 21. http://vcresearch.berkeley.edu/news/campus-dedicates-li-ka-shing-cen ter-biomedical-and-health-sciences-philanthropist-receives-berk.

Roepstorff, Andreas. 2001. "Brains in Scanners: An Umwelt of Cognitive Neuroscience." *Semiotica* 134, no. 1/4: 747–765.

Rorschach, Hermann. 1921. *Psychodiagnostik: Psychodiagnostics. Tafeln. Plates.* New York: Hans Huber.

Rorschach, Hermann. 2008. *Psychodiagnostics: A Diagnostic Test Based on Perception.* Vancouver: Read Books.

Rose, Nikolas. 2004. "The Neurochemical Self and Its Anomalies." In *Risk and Morality*, ed. Richard Victor Ericson and Aaron Doyle, 407–437. Toronto: University of Toronto Press.

Rose, Nikolas. 2007. *The Politics of Life Itself: Biomedicine, Power, and Subjectivity in the Twenty-First Century*. Princeton, NJ: Princeton University Press.

Rose, Nikolas, and Joelle M. Abi-Rached. 2013. *Neuro: The New Brain Sciences and the Management of the Mind*. Princeton, NJ: Princeton University Press.

Rosenberg, Charles E. 1997. *No Other Gods: On Science and American Social Thought*. Baltimore: Johns Hopkins University Press.

Ross, Andrew S. 2013. "Rock Health Moving to Mission Bay." *SFGate*, June 4. http://www.sfgate.com/business/bottomline/article/Rock-Health-moving-to-Mission-Bay-4577388.php#src=fb.

Rouse, Caroline. 2009. *Uncertain Suffering: Racial Health Care Disparities and Sickle Cell Disease*. Berkeley: University of California Press.

Ruesch, Jürgen K., and Gregory Bateson. 1951. *Communication: The Social Matrix of Psychiatry*. New York: Norton.

Said, Carolyn. 2011. "Pfizer to Open Research Center in SF's Mission Bay." *SFGate*, July 22. http://www.sfgate.com/business/article/ Pfizer-to-open-research-center-in-SF-s-Mission-Bay-2353818.php#src=fb.

Sandel, Michael J. 2013. *What Money Can't Buy: The Moral Limits of Markets*. New York: Farrar, Straus and Giroux.

Savard, Jacqueline. 2013. "Personalised Medicine: A Critique on the Future of Health Care." *Journal of Bioethical Inquiry*, 1, no. 2: 197–203. http://dx.doi.org/ 10.1007/s11673-013-9429-.

Saxenian, Anna Lee. 1996. *Regional Advantage: Culture and Competition in Silicon Valley and Route 128*. Cambridge, MA: Harvard University Press.

Schlosser, Allison V., and Kristi Ninnemann. 2012. "Introduction to the Special Section: The Anthropology of Psychopharmaceuticals: Cultural and Pharmacological Efficacies in Context." *Culture, Medicine, and Psychiatry*, 36, no. 1: 2–9. http://dx.doi.org/10.1007/s11013-012-9249-z.

Schrecker, E. 2010. *The Lost Soul of Higher Education: Corporatization, the Assault on Academic Freedom, and the End of the American University*. New York: New Press.

Schüll, Natasha Dow. 2010. *Machine Life: Design and Dependency in Las Vegas*. Princeton, NJ: Princeton University Press.

Shankar, Shalini. 2008. *Desi Land: Teen Culture, Class, and Success in Silicon Valley*. Durham, NC: Duke University Press.

Shapin, Steven. 2008. *The Scientific Life*. Chicago: University of Chicago Press.

Sharp, Lesley A. 2000. "The Commodification of the Body and Its Parts." *Annual Review of Anthropology*, 29:287–328.

Sharp, Lesley A. 2011. "The Invisible Woman: The Bioaesthetics of Engineered Bodies." *Body & Society*, 17, no. 1: 1–30. http://dx.doi.org/10.1177/1357034X10394667.

Sharp, Lesley A. 2013. *The Transplant Imaginary: Mechanical Hearts, Animal Parts, and Moral Thinking in Highly Experimental Science.* Berkeley: University of California Press.

Shore, Chris, and Susan Wright. 1999. "Audit Culture and Anthropology: Neo-Liberalism in British Higher Education." *Journal of the Royal Anthropological Institute,* 5, no. 4: 557–575.

Shumar, Wesley. 1997. *College for Sale: A Critique of the Commodification of Higher Education.* London: Falmer Press.

Shweder, Richard A. 1991. *Thinking through Cultures: Expeditions in Cultural Psychology.* Cambridge, MA: Harvard University Press.

Simmel, Georg. 1909. "Bridge and Door." In *Rethinking Architecture: A Reader in Cultural Theory,* ed. Neil Leach, 66–69. London: Routledge.

Sismondo, Sergio. 2009. "Ghosts in the Machine: Publication Planning in the Medical Sciences." *Social Studies of Science,* 39, no. 2: 171–198. http://dx.doi.org/10.1177/0306312708101047.

Skripka-Serry, Julia. 2013. "The Great Neuro-Pipeline Brain Drain (and Why Big Pharma Hasn't Given Up on CNS Disorders)." *Drug Discovery World,* Fall. http://www.ddw-online.com/p-216813.

Slaughter, Sheila, and Gary Rhoades. 2004. *Academic Capitalism and the New Economy: Markets, State, and Higher Education.* Baltimore: Johns Hopkins University Press.

Smith, Kerri. 2011. "Trillion-Dollar Brain Drain." *Nature News,* 478, no. 7367: 15. http://dx.doi.org/10.1038/478015a.

Solomon, Miriam. 2015. *Making Medical Knowledge.* Oxford University Press.

Solomon, Mildred Z. 2010. "The Ethical Urgency of Advancing Implementation Science." *American Journal of Bioethics,* 10, no. 8: 31–32. http://dx.doi.org/10.1080/15265161.2010.494230.

Solomon, N. 1993. "Laboratory Innovations." *Architecture,* 82, no. 3: 123–127.

Spratto, George R., and Adrienne L. Woods. 2013. *Delmar Nurse's Drug Handbook 2013.* Albany, NY: Delmar-Cengage Learning.

Star, Susan Leigh. 1989. *Regions of the Mind: Brain Research and the Quest for Scientific Certainty.* Stanford, CA: Stanford University Press.

Star, Susan Leigh. 1995. *Ecologies of Knowledge: Work and Politics in Science and Technology.* Albany: SUNY Press.

Star, Susan Leigh, and Karen Ruhleder. 1996. "Steps toward an Ecology of Infrastructure: Design and Access for Large Information Spaces." *Information Systems Research,* 7, no. 1: 111–134. https://doi.org/10.1287/isre.7.1.111.

Star, Susan Leigh, and James R. Griesemer. 1999. "Institutional Ecology, 'Translation,' and Bounday Objects. Amateurs and Professionals in Berkeley's Museum of Vertebrate Zoology 1907–39." In *The Science Studies Reader*. New York: Routledge.

Starr, Paul. 1988. "The Meaning of Privatization." *Yale Law and Policy Review*, 6:6–41.

Stewart, Thomas A. 1997. *Intellectual Capital: The New Wealth of Organizations*. New York: Doubleday Currency.

Stone, Glen Davis. 2010. "The Anthropology of Genetically Modified Crops." *Annual Review of Anthropology*, 39, no. 1: 381–400. https://www.annualreviews.org/doi/abs/10.1146/annurev.anthro.012809.105058.

Stowall, Sten. 2011. "R&D Cuts Curb Brain-Drug Pipeline." *Wall Street Journal*, March 27. http://online.wsj.com/article/ SB10001424052748704474804576222463 927753954.html.

Strathern, Marilyn. 1995. "Displacing Knowledge: Technology and the Consequences for Kinship." In *Conceiving the New World Order: The Global Politics of Reproduction*, ed. Faye D. Ginsburg and Rayna Rapp, 346–364. Berkeley: University of California Press.

Strathern, Marilyn. 2000. *Audit Cultures: Anthropological Studies in Accountability, Ethics and the Academy*. London: Routledge.

Styhre, Alexander. 2015. *Financing Life Science Innovation: Venture Capital, Corporate Governance and Commercialization*. London: Palgrave Macmillan.

Sugarman, Jeremy, and W. Gillies McKenna. 2003. "Ethical Hurdles for Translational Research." *Radiation Research*, 160, no. 1: 1–4.

Sunder Rajan, Kaushik. 2006. *Biocapital: The Constitution of Postgenomic Life*. Durham, NC: Duke University Press.

Sunder Rajan, Kaushik, ed. 2012. *Lively Capital: Biotechnologies, Ethics, and Governance in Global Markets*. Durham, NC: Duke University Press.

Sunder Rajan, Kaushik. 2017. *Pharmocracy: Value, Politics, and Knowledge in Global Biomedicine*. Durham, NC: Duke University Press Books.

Sunder Rajan, Kaushik, and Sabina Leonelli. 2013. "Introduction: Biomedical Transactions, Postgenomics, and Knowledge/Value." *Public Culture*, 25, no. 3 (71): 463–475. http://dx.doi.org/10.1215/08992363-2144607.

Theodore, William H., and Robert S. Fisher. 2004. "Brain Stimulation for Epilepsy." *The Lancet Neurology*, 3, no. 2: 111–118.

Thompson, Jennifer Jo, Cheryl Ritenbaugh, and Mark Nichter. 2009. "Reconsidering the Placebo Response from a Broad Anthropological Perspective." *Culture, Medicine and Psychiatry*, 33, no. 1: 112–152. http://dx.doi.org/10.1007/s11013-008-9122-2.

Throsby, Karen. 2009. "The War on Obesity as a Moral Project: Weight Loss Drugs, Obesity Surgery and Negotiating Failure." *Science as Culture*, 18, no. 2: 201. http://dx.doi.org/10.1080/09505430902885581.

Timmerman, Luke. 2010. "The Mission Bay Biotech Cluster: Antibodies, RNAi, Biofuels, & More." *Xconomy*, June 29. http://www.xconomy.com/san-francisco/2010/06/29/the-mission-bay-biotech-cluster-antibodies-rnai-biofuels-more/.

Traweek, Sharon. 1988. *Beamtimes and Lifetimes*. Cambridge, MA: Harvard University Press.

Tsing, Anna Lowenhaupt. 2005. *Friction: An Ethnography of Global Connection*. Princeton, NJ: Princeton University Press.

Tutton, Richard. 2011. "Promising Pessimism: Reading the Futures to Be Avoided in Biotech." *Social Studies of Science*, 41, no. 3: 411–429. http://dx.doi.org/10.1177/0306312710397398.

Tyfield, David. 2011. *The Economics of Science: A Critical Realist Overview, Volume 1: Illustrations and Philosophical Preliminaries*. London: Routledge.

Tyfield, David, Rebecca Lave, Samuel Randalls, and Charles Thorpe, eds. 2017. *The Routledge Handbook of the Political Economy of Science*. New York: Routledge.

Van der Geest, Sjaak, Susan Reynolds Whyte, and Anita Hardon. 1996. "The Anthropology of Pharmaceuticals: A Biographical Approach." *Annual Review of Anthropology*, 25, no. 1: 153–178.

Vanderheiden, Gregg, and Jim Tobias. 2000. "Universal Design of Consumer Products: Current Industry Practice and Perceptions." *Proceedings of the Human Factors and Ergonomics Society Annual Meeting*, 44, no. 32: 6–21.

Van Gerven, Joop, and Adam Cohen. 2011. "Vanishing Clinical Psychopharmacology." *British Journal of Clinical Pharmacology*, 72, no. 1: 1–5. http://dx.doi.org/10.1111/j.1365-2125.2011.04021.x.

Veblen, Thorstein. 1918. *The Higher Learning in America: A Memorandum on the Conduct of Universities by Business Men*. New York: B. W. Huebsch.

Venkatesan, Soumhya, Laura Bear, Penny Harvey, Sian Lazar, Laura Rival, and AbdouMaliq Simone. 2018. "Attention to Infrastructure Offers a Welcome Reconfiguration of Anthropological Approaches to the Political." *Critique of Anthropology*, 38, no. 1: 3–52. https://doi.org/10.1177/0308275X16683023.

Venturi, Robert. 1999. "Thoughts on the Architecture of the Scientific Workplace: Community, Change, and Continuity." In *The Architecture of Science*, ed. Peter Galison and Emily Thompson, 385–398. Cambridge, MA: MIT Press.

Vinluan, Frank. 2012. "GSK's Drug Pipeline: DPUs Produce More Drug Candidates." *MedCity News*, February 7. http://medcitynews.com/2012/02/a-tale-of-two-pharmas -gsks-pipeline-shows-promise-azns-prospects-dim/.

Vrecko, Scott. 2010. "Neuroscience, Power and Culture: An Introduction." *History of the Human Sciences*, 23, no. 1: 1–10. http://dx.doi.org/10.1177/0952695109354395.

Wailoo, Keith. 2017. *How Cancer Crossed the Color Line*. Oxford: Oxford University Press. (Orig. pub. 2011.)

Wailoo, Keith, Julie Livingston, Steven Epstein, and Aaron Aronowiz. 2010. *Three Shots at Prevention: The HPV Vaccine and the Politics of Medicine's Simple Solutions*. Baltimore: Johns Hopkins University Press.

Walsh, James. 2013a. "Medtronic Device Will Collect Data as It Treats Brain Disorders." *Medtronic Union Bulletin*, August 10. http://union-bulletin.com/news/2013/ aug/10/medtronic-device-will-collect-data-it-treats-brain/.

Walsh, James. 2013b. "Medtronic: Overtightening Screw Can Damage Wire That Delivers Therapy to Brain." *Star Tribune*, May 7. http://www.startribune.com/business/206049721.html.

Wang, Andrew L. 2012. "Northwestern University Leads Nation in Tech Transfer Revenue." *Crain's Chicago Business*, October 29. http://www.chicagobusiness .com/article/20121027/ISSUE01/310279974/northwestern-university-leads-nation -in-tech-transfer-revenue.

Warner, John Harley. 2014. *The Therapeutic Perspective: Medical Practice, Knowledge, and Identity in America, 1820-1885*. (With a new preface by the author.) Princeton: Princeton University Press.

Watch, Daniel. 2010. "Trends in Lab Design." *WBDG: Whole Building Design Guide*. http://www.wbdg.org/resources/labtrends.php.

Wehling, Martin. 2008. "Translational Medicine: Science or Wishful Thinking?" *Journal of Translational Medicine* 6:31. https://doi.org/10.1186/1479-5876-6-31.

Wehling, Martin, ed. 2015. *Principles of Translational Science in Medicine: From Bench to Bedside*. Cambridge, MA: Academic Press.

Welz, Gisela. 2003. "The Cultural Swirl: Anthropological Perspectives on Innovation." *Global Networks*, 3, no. 3: 255–270. http://dx.doi.org/10.1111/1471-0374.00061.

West, Harry G., and Todd Sanders, eds. 2003. *Transparency and Conspiracy: Ethnographies of Suspicion in the New World Order*. Durham, NC: Duke University Press.

Whitehouse, Peter J., Konrad Maurer, and Jesse F. Ballenger, eds. 2000. *Concepts of Alzheimer Disease: Biological, Clinical, and Cultural Perspectives*. Baltimore: Johns Hopkins University Press.

Whyte, Susan Reynolds, Sjaak van der Geest, and Anita Hardon. 2003. *Social Lives of Medicines*. Cambridge: Cambridge University Press.

Whyte, Susan Reynolds, Michael H. Whyte, Lotte Meinert, and Betty Kyaddondo. 2006. "Treating AIDS: Dilemmas of Unequal Access in Uganda." In *Global Pharmaceuticals: Ethics, Markets, Practices*, ed. Adriana Petryna, Andrew Lakoff, and Arthur Kleinman, 240–262. Durham, NC: Duke University Press.

Wolfe, Audra. 2002. "Federal Policy Making for Biotechnology, Executive Branch, ELSI." *Encyclopedia of Ethical, Legal, and Policy Issues in Biotechnology*. Hoboken, NJ: Wiley.

Woolgar, Steve, Cateline Coopmans, and Daniel Neyland. 2009. "Does STS Mean Business?" *Organization*, 16, no. 1: 5–30. http://dx.doi.org/10.1177/1350508408098983.

Wright, Peter, and Andrew Treacher. 1982. *The Problem of Medical Knowledge*. Edinburgh: Edinburgh University Press.

Wu, Bernhard S., and Michael A. Reslinski. 2016. "The Value of Royalty." *Nature Biotechnology*, 34, no. 7: nbt.3624. https://doi.org/10.1038/nbt.3624.

Wynne, Brian. 1993. "Public Uptake of Science: A Case for Institutional Reflexivity." *Public Understanding of Science*, 2, no. 4: 321–337. http://dx.doi.org/10.1088/0963-6625/2/4/003.

Wynne, Brian, Jack Stilgoe, and James Wilsdon. 2005. *The Public Value of Science: Or How to Ensure That Science Really Matters*. London: Demos.

Wynne, Irwin Alan, and Brian Wynne, eds. 1996. *Misunderstanding Science? The Public Reconstruction of Science and Technology*. Cambridge: Cambridge University Press.

Young, Allan. 1997. *The Harmony of Illusions: Inventing Post-Traumatic Stress Disorder*. Princeton, NJ: Princeton University Press.

Young, Allan. 2007. "America's Transient Mental Illness." In *Subjectivity: Ethnographic Investigations*, ed. João Biehl, Byron J. Good, and Arthur Kleinman. Berkeley: University of California Press.

Zeng, David, Dennis R. Hiltunen, and Majid Manzari, eds. 2008. *Geotechnical Earthquake Engineering and Soil Dynamics IV*. Reston, VA: American Society of Civil Engineering.

Zerhouni, Elias A. 2005. "Translational and Clinical Science: Time for a New Vision." *New England Journal of Medicine*, 353, no. 15: 1621–1623.

Zerubavel, Eviatar. 1999. *Social Mindscapes: An Invitation to Cognitive Sociology*. Cambridge, MA: Harvard University Press.

Zlolniski, Christian. 2006. *Janitors, Street Vendors, and Activists: The Lives of Mexican Immigrants in Silicon Valley*. Berkeley: University of California Press.

# Index

21st Century Cures Act, 22

Abi-Rached, Joelle M., 20
Academy of Translational Medicine, 35
Actor network theory (ANT), 18, 115,
    180
Ahlburg, Dennis, 139
AIDS crisis, 43, 216–217
Althusser, Louis, 187
Altruism, 166–168, 177
Alzheimer's
    anthropological constructions and, 126
    Carlisle and, 125–130, 137, 140
    dementia and, 126, 197–198, 207
    financialization and, 49, 52, 54, 65,
        83, 86
    goals in mind and, 127–130
    innovation and, 149, 197
    laboratories and, 125–133
    Lock and, 83
    neuronal death and, 127
    normal aging and, 125
    Pfizer and, vii
    research and development (R&D) on,
        125–133
    translational neuroscience (TN) and,
        10–11, 231
America Invents Act, 22
Anthropology, 222
    Boellstorff and, 117
    cognition and, 17, 19, 32, 119

Douglas and, 17, 57, 217
economic, 24, 32
Fassin and, 216
Fischer and, 205, 218
Grosof and, 85
Hess and, 18, 146
Ho and, 32
innovation and, 151, 158
laboratories and, 99, 110, 117, 119–
    120, 123, 126, 138, 146
metaepistemic space and, 57
political economy and, 32, 47
psychiatry and, 79–80
science and, 6, 13, 16–17, 47, 57, 79,
    85, 96, 120, 138, 205, 218, 244n12
translational neuroscience (TN) and,
    6, 13, 16–19, 24, 26, 96, 151
Applbaum, Kalman, 88
Apple, 55, 60, 62f, 150
Architecture
    collaboration and, 19
    commensurability and, 235
    commercial, 19, 30, 48, 108
    durability and, 30
    epistemologies and, 97, 101, 108–109,
        121, 124–125, 130–142, 180
    fantasies and, 130–132, 235
    glass and, 102, 109, 123–125, 132–
        141, 163–165, 246n11, 247n14
    laboratories and, 97, 101, 108–109,
        121, 124–125, 130–142, 246nn11,12

Architecture (cont.)
  Miu and, 135–139
  research and development (R&D) and,
    97, 101, 108–109, 121, 124–125,
    130–142
  transparency and, 102, 109, 123–125,
    130–142, 163–165, 246n11, 247n14
AstraZeneca, 54–56, 66, 98, 111–112,
  245n6
Atlanta Clinical and Translational
  Science Institute (Atlanta-CTSI), 35
Autism, 38, 43, 83

Baudrillard, Jean, 80
Bay City Capital, 88
Bayer, 102
Bayh-Dole Act, 22, 25, 37, 67, 244n6
"Bedside to bench" approach, 204–207
Benches. See Laboratories
"Bench to bedside" approach, 39–40,
  43, 71, 84, 128, 131, 204–205, 207
Bernstein, Mark, 213
Biehl, João, 27
Biological duress, 193
Biological material
  changing semiotics of, 182
  clinical interpretation and, 184
  concept of, 41
  discourses for, 148, 187–195
  disorder matching and, 40–41, 119,
    209
  existing therapeutics and, 185–186
  innovation and, 148, 159, 182–195,
    199
  micropractices and, 15
  patients and, 218
  productivity and, 187–188
Biomarkers, 41, 69, 71, 75, 80–82, 191,
  201, 234
Biotechnology
  biopharmaceuticals and, 7, 15, 24 (see
    also Pharmaceuticals)

financialization and, 59, 67, 76, 80,
  82, 86, 89, 92–94
innovation and, 145–154, 160–161,
  165–169, 182, 186, 188, 192, 195
laboratories and, 98–104, 110, 117,
  123
neuropharmacoepistemology and,
  233–234
partnerships and, 7
patients and, 207, 213–220
political economy and, 40–41
translational neuroscience (TN) and,
  1–7, 14–15, 19, 22–23, 26, 241n2
Birch, Kean, 23
Blood tests, 184–186, 208
Boellstorff, Tom, 117
Booz & Company, 56, 62f, 63, 120
Boston Scientific, 183
Boston University, 242n5
Brain science, 231–233
  financialization and, 84
  innovation and, 149, 154, 192
  political economy and, 38, 40
  translational neuroscience and, 3–4,
    10, 13, 20–21, 26
Bristol-Myers Squibb, 56
Building an External Innovation Capability
  (Booz & Company), 60

Cancer, 10, 16, 43, 72, 74, 86, 158, 160,
  186, 205
Capitalism
  financialization and, 8, 15, 22–23, 43,
    64, 92–93, 95, 137, 242n12
  life sciences and, 23
  venture capital and, 2, 5, 38 (see also
    Venture capital)
Carlisle, Tom, 125–130, 137, 140
Center for Biomedical Engineering
  Innovation and Design, 57
Center for Clinical and Translational
  Science, 35

Center for Health and Biomedical
    Innovation, 135
Center for Science, Technology,
    Medicine & Society, 85
Centers for Therapeutic Innovation
    (CTI), 66, 104–108, 138, 245n5
Central nervous system (CNS) sector
    abandonment of, 53–55, 57, 69–72,
        86, 156
    financialization and, 53–57, 64–65,
        69–71, 76, 84–89, 95
    innovation and, 153, 169
    research and development (R&D) and,
        47, 53–57, 64–65, 69–72, 76, 84–89,
        95, 153, 156, 169
    stagnation in, 47
    translational neuroscience (TN) and,
        7, 10–13, 23, 26, 241n2
Chen, John, 214
Chiang, Simon, 169, 175, 183
China Basin, 104, 109
Cleveland Clinic, 14, 204, 210–213,
    223, 226, 248n2
Clientalization, 179–180
Clinical and Translational Science
    Awards (CTSAs), 34, 36, 89–90, 114f
Clinical and Translational Science
    Institute (CTSI)
    center of, 110
    cognitive artifacts and, 119–120
    collaboration and, 109–121
    contoured actions and, 121–122
    epistemic action and, 119
    extensional format and, 118–119
    as interactive portal, 110–111, 120
    partnerships and, 109–111, 118
    research and development (R&D) and,
        109–122, 138, 142, 220, 245n5
    risk and, 220
Clinical trials
    financialization and, 58, 61, 71, 81,
        85–86, 88

    innovation and, 159, 170, 185, 189,
        192, 194
    laboratories and, 104, 142
    patients and, 215–216, 219–221
    safety and, 10–12, 58, 61, 71, 81, 85–
        86, 88, 104, 142, 159, 170, 185, 189,
        192, 194, 215–216, 219–221
    translational neuroscience (TN) and,
        10–12
Cognition
    anthropology and, 17, 19, 32, 119
    financialization and, 54, 76–78, 87
    innovation and, 198–199
    laboratories and, 119–120, 245n8
    mice and, 198
    political economy and, 38, 41
    realization and, 244n7
    revolution in study of, 41, 54
    translational neuroscience (TN) and,
        17, 19, 26
Cognitive artifacts, 119–120
CoLaborator, 102
Collaboration
    architecture and, 19
    Clinical and Translational Science
        Institute (CTSI) and, 109–121
    commercialism and, 19, 59, 90, 181, 220
    expertise mining and, 113–117
    financialization and, 53, 55, 59–63,
        69, 90
    information engines and, 120
    innovation and, 172, 175, 178, 181–
        182, 192
    laboratories and, 102, 104–121, 125,
        136–141
    Neurotechnology Investing and
        Partnering Conferences and,
        146–148
    political economy and, 34–35, 38
    Post-Fordist models and, 115–116
    translational neuroscience (TN) and,
        17, 19, 236

Collaboration (cont.)
  transparency and, 132–142, 247n14
  virtualization and, 55, 111–112
Commensurability, 235–236, 246n10
Commercialization
  architecture and, 19, 30, 48, 108
  Bayh-Dole and, 244n6
  collaboration and, 19, 59, 90, 181, 220
  innovation and, 10, 21, 46, 48, 61,
    107, 146, 180–183, 244n6
  intellectual property and, 90–91
  laboratories and, 98, 107–108, 119–
    120, 123, 245n8
  leveraging and, 46, 89, 91, 118
  NeuroInsights and, 2, 6–7, 23,
    146–148
  partnerships and, 3, 61, 67, 104, 108,
    245n8
  patients and, 214, 219
  pharmaceuticals and, 2–3, 46, 65, 89,
    91
  political economy and, 36, 44, 46, 48
  privatization and, 61
  translational neuroscience (TN) and,
    2–3, 7, 9–11, 21–25, 91, 245n8
  universities and, 2–3, 11, 21–22, 50,
    65, 67, 89–92, 98, 120, 123, 181,
    245n8
  US National Institutes of Health (NIH)
    and, vii, 2–3, 36, 50, 105
Constructivism, 41, 193, 217–219
Cornell University, 66, 81
Critical neuroscience, 20

D'Andrade, Roy, 119
Davies, Gail, 199, 218–222
Deal flow, 117, 152–153
Decade of the brain, 13, 40
Deep brain stimulation (DBS) devices,
    159, 211–214, 217–219, 248n5,
    249n11
Defunding, 92, 232, 245n15
Dementia, 126, 197–198, 207

Depression
  DBS devices and, 211
  financialization and, 49, 54, 80, 82–83
  innovation and, 184–185
  patients and, 211
  screening tests for, 184–185
  translational neuroscience (TN) and,
    10, 231
De-risking mechanisms, 67, 69, 95
Design Process as a Critical Component of
    the Anthropology of Technology, The
    (Kingery), 138
Destigmatization, 76, 244n8
Detethering, 236–239
Diagnostic and Statistical Manual (DSM),
    83
Diagnostics
  financialization and, 82, 86, 89
  innovation and, 146, 149, 159, 175,
    178, 183–188, 191, 195
  integration and, 236
  patients and, 208–209, 222
  political economy and, 36, 38–40, 46
  risk and, 183–184
  translational neuroscience (TN) and,
    2, 5, 12, 16, 27
Discovering New Therapeutic Uses for
    Existing Molecules (NIH), vii, 56,
    105, 107
Douglas, Mary, 17, 57, 217
Downsizing, 54, 56–57, 69, 231
Down syndrome, 30
Drosophila melanogaster, 30–33, 38,
    218, 241nn1,4
Drugs for Life (Dumit), 184–185
Duke University, 39
Dumit, Joseph, 82–83, 184–185, 209
Durkheim, Émile, 17, 85

Economic bioethics, 239
Eli Lilly, 56
Elitism, 31, 139
Elliot, Leonard, 166–169, 177–178

Emory University, 35

"Ensuring Public Trust: Conflicts of Interest in Psychiatry" (Insel), 73

Entrepreneurs
  financialization and, 52, 59, 85, 89, 91–92
  innovation and, 147–148, 151, 166–173, 176–177, 181, 188–189, 247n1
  investment and, 1–2, 6–7, 13–14, 52, 58–59, 119, 143, 145, 147, 151, 166, 168, 176–177, 181, 188–189, 232
  laboratories and, 99, 112–113, 116, 119, 122, 139–140, 143
  *Neurotechnology Industry Report* and, 6, 23, 182
  political economy and, 34, 38
  research and development (R&D) and, 59, 169, 232
  startups and, 6, 52, 92, 99, 102, 110, 151, 166, 168, 173, 188–189
  translational neuroscience (TN) and, 1–2, 6–7, 13–14, 22–23
  universities and, 89–93

Epilepsy
  financialization and, 68
  innovation and, 153, 177, 195–196
  patients and, 209–214, 217–219, 223–225, 248n4
  translational neuroscience (TN) and, 10

Epistemic action, 119

Epistemologies, 230, 235
  architecture and, 97, 101, 108–109, 121, 124–125, 130–142, 180
  cognitive artifacts and, 119–120
  constructivism and, 41, 193, 217–219
  ethics and, 247n4
  feminist, 205–206
  financialization and, 57–58, 78, 83
  functionability and, 80
  imaginaries and, 10, 132

  importance of design and, 97–99
  innovation and, 145, 157–158, 180–181, 188, 190, 193, 198–199
  laboratories and, 17–18, 97–99, 103, 107, 119, 130, 137–139
  legibility of facts and, 132
  manipulability and, 80, 194
  mechanism of action (MoA) and, 157–158, 188, 190
  metaepistemic space and, 47, 57
  neuropharmacoepistemology and, 233–234
  ontological issues and, 26, 83, 185–186, 195, 200, 243n5, 248n10
  patients and, 205
  pharmaceuticals and, 237
  proximity/intimacy and, 103–109
  role of connections and, 119
  translational neuroscience (TN) and, 9, 15, 27, 232, 236
  translational thinking and, 233–234

Epistemology of parts, 27, 145, 198–199, 235

Epstein, Steve, 203

Ethical, Legal, Social Implications (ELSI) funding, 226

Ethics
  ethnographic analysis and, 1–2, 14–16, 19–20, 26
  financialization and, 66, 73–78, 94, 236–239
  Greely on, 1
  innovation and, 195–201, 247n4
  laboratories and, 98, 131, 142
  Lynch on, 1
  neuroenhancement and, 1
  partial subjects and, 195–201
  patients and, 204, 213, 216, 219, 226–227, 229
  political economy and, 32
  prescribing practices and, 74–80
  translational neuroscience (TN) and, 1–2, 14–16, 19–20, 26

Ethnographic analysis
  inextricability and, 8, 17
  laboratories and, 16–19
  moral imperatives and, 9–13
  spaces of investigation and, 13–14
  translational neuroscience (TN) and,
    1–27
Etzkowitz, Henry, 140
European Advanced Translational
  Research Infrastructure in Medicine
  (EATRIS), 35
European College of
  Neuropsychopharmacology, 70
Expertise mining, 113–117, 120
Externalization
  financialization and, 55–64, 69–70,
    86, 89–91, 93–94, 231
  innovation and, 60–64
  laboratories and, 106–107, 111
  risk and, 60–64, 230–232, 237
  translational neuroscience (TN) and,
    15, 25
  universities and, 60–64

Farquhar, Judith, 247n4
Fassin, Didier, 216
Fast tracking, 216
Federal funding, 7, 23, 25, 31, 33, 37,
  41, 65, 78, 217, 221, 226, 245n15
Federal Technology Transfer Act, 37
Feminism, 205–206
Feyerabend, Paul, 235
Financialization, x
  academic livelihoods and, 226–227
  Alzheimer's and, 49, 52, 54, 65, 83,
    86
  biotechnology and, 59, 67, 76, 80, 82,
    86, 89, 92–94
  capitalism and, 8, 15, 22–23, 43, 64,
    92–93, 95, 137, 242n12
  central nervous system (CNS) research
    and, 53–57, 64–65, 69–71, 76, 84–
    89, 95

clinical trials and, 58, 61, 71, 81, 85–
  86, 88
cognition studies and, 54, 76–78, 87
collaboration and, 53, 55, 59–63, 69,
  90
commercialization and, 2–3 (see also
  Commercialization)
deal flow and, 117, 152–153
depression and, 49, 54, 80, 82–83
detethering and, 236–239
diagnostics and, 82, 86, 89
entrepreneurs and, 52, 59, 85, 89,
  91–92
epilepsy and, 68
epistemologies and, 57–58, 78, 83
ethics and, 66, 73–78, 94, 236–239
externalization and, 55–64, 69–70, 86,
  89–91, 93–94, 231
free market and, 110
funding and, 52, 56–60, 62–63, 65, 78,
  92, 244n9
global economic crisis and, viii,
  14–15, 22, 33–34, 48, 57, 80, 86,
  89, 94
Humanitarian Device Exemption
  (HDE) and, 215–220
increase of, 230–232
intellectual property (IP) and, 58, 62f,
  75, 77, 91–92
investment and, 49–53, 57–59, 62–64,
  67–69, 80, 85–86, 93
knowledge problems and, 84–89
leveraging and, 14, 26, 46, 49–50, 55–
  56, 60–63, 89, 91, 95, 118, 168, 186–
  187, 209, 230
licensing and, 22, 36, 52–53, 58–63,
  66–69, 92, 102, 106, 140, 147, 161,
  166, 242n10, 244n6
lobbying and, 2, 6–7, 11
market issues and, 49–54, 60–68, 71–
  72, 75–76, 86–88, 91–96, 245n15
mechanism of action (MoA) and, 70,
  81

morals and, 78, 188, 196, 227
neoliberalism and, 22–25, 56, 61, 66, 89, 92–93, 95
neurotechnology and, 49–54, 59, 65, 67, 75, 77, 85–89, 96
*Neurotechnology Industry Report* and, 6, 23, 182
Neurotechnology Investing and Partnering Conferences and, 2, 5, 15, 21, 42, 52, 54, 65, 88–89, 98, 132, 146–149, 163, 173, 176–177, 181, 183, 188, 195, 201, 203, 232, 234, 236
opportunity and, 73–80
partnerships and, 3, 49, 52–69, 85, 90–91, 93, 104, 108, 230–231, 236, 238, 242n10, 245nn8,15
patents and, 22, 24, 36, 38, 51, 59, 61, 65, 72–77, 81, 85, 92, 140, 154, 172, 175, 181, 196, 234, 247n13
Pfizer and, 56, 60, 66, 68
pharmaceuticals and, 49–61, 64–95, 238
pipelines and, 52, 58, 60–61, 66–67, 70–71, 80–81, 85
privatization and, 21–22, 25, 60–61, 91–93, 95, 243n3, 245n15
profit and, 51, 56, 64–69, 74, 81, 86–88
psychiatry and, 52–55, 66, 70–81, 88, 182, 244nn8,9
research and development (R&D) and, 49, 55–64, 69–70, 85, 88–89, 93–96, 238, 243n2
risk and, 49–53, 57, 59–71, 75, 80, 83–95
startups and, 50–52, 62f, 85, 92
taxpayers and, 30–33
translational neuroscience (TN) and, 14, 22–26, 49, 51, 55–57, 59, 66–70, 74, 77–78, 81–82, 85, 91, 93–96, 230–232
understanding TSM and, 230–232

US National Institutes of Health (NIH) and, 34–36, 50, 56, 64, 66, 90, 105, 217
venture capital and, 2, 5, 38, 46, 50–51, 62f, 88, 99–102, 110, 117–118, 120, 142, 146–147, 150–155, 166, 175, 181–183, 190, 204, 247n3
vertical disintegration and, 93–96
Financial mechanisms, 67
Fischer, Michael M. J., 205, 218
FitzGerald, Garret, 43–44
fMRI, 13, 20, 40, 163, 196, 208
Fournier, Jay C., 72
Fox, Howard S., 41
Franklin, Sarah, 194
Free market, 110
Friedman, Richard A., 66–67
"Frontiers in Neurotechnology" event, 184
Fruit flies, 30–33, 38, 218, 241nn1,4
Functionability, 80
Funding
academic livelihoods and, 226–227
defunding and, 92, 232, 245n15
ELSI, 226
expertise mining and, 116–117
federal, 7, 23, 25, 31, 33, 37, 41, 65, 78, 217, 221, 226, 245n15
financialization and, 52, 62–63, 65, 78, 92, 244n9
grants and, 61–63, 125, 215, 249n14
innovation and, 56–60, 148, 188, 207
investment and, 7, 11, 33, 37, 43, 52, 62–63, 188
laboratories and, 41, 104–105, 116, 120, 125–126
partnerships and, 56–60
pharmaceuticals and, 56–60
philanthropy and, 43
political economy and, 29–37, 41, 43, 46–47

Funding (cont.)
　research and development (R&D) and,
　　7, 11, 25, 29–37, 41, 43, 62–63, 78,
　　92, 104–105, 116, 120, 125–126,
　　207, 217, 226, 232, 244n9, 245n15
　scholarships and, 61–63
　Technology Accelerator Fund and,
　　57–58
　translational neuroscience (TN) and,
　　7, 11, 23, 25, 231–232
　universities and, 7, 11, 29, 34–35, 37,
　　41, 47, 65, 92, 104–105, 116, 120,
　　125–126, 245n15
　venture capital and, 2, 5, 38, 46, 50,
　　88, 99–102, 110, 117–118, 120, 142,
　　146–147, 150–155, 166, 175, 181–
　　183, 190, 204, 247n3
Futurism, 4–6, 9, 12

Geertz, Clifford, 176–180
Genentech, 101, 167
General Electric, 197
Georgia Institute of Technology, 35
Gieryn, Thomas, 137, 139
"Give Me a Laboratory and I Will Raise
　the World" (Latour), 129
GlaxoSmithKline, 54, 56
Global economic crisis, viii, 14–15, 22,
　33–34, 48, 57, 80, 86, 89, 94
Global health, 16–19, 55, 74, 83, 98, 227
Globalization, 35–36, 54, 56, 98, 137,
　140
Goals in mind, 127–130
Good science, 51, 64–65, 75, 153
Google, 55, 60, 62f
Grants, 61–63, 125, 215, 249n14
Greely, Hank, 1
Grosof, David, 85–86
Guardian, The, 158

Hall, Michael, 50–51, 117–118, 151–
　154, 165–169, 175–178, 180
Haraway, Donna, 246n9

Harding, Sandra, 246n9
Harvard, 21, 52, 151, 169, 175, 205, 223
Helmreich, Stefan, 247n1
Hess, David, 18, 146
HIV, 43
Ho, Karen, 32
Holborn, Mary, 161
Humanitarian Device Exemption (HDE),
　215–220
Hutchins, Edwin, 119

Imaginaries, 10, 132
IMED Biotech, 55, 66, 111
Implants, 7, 159, 186, 205, 211, 213
Inextricability, 8–9, 246n9
Information engines, 120
Initial public offerings (IPOs), 152
Innovation
　21st Century Cures Act and, 22
　Alzheimer's and, 149, 197
　America Invents Act and, 22
　anthropology and, 151, 158
　Bayh-Dole Act and, 22, 25, 37, 67,
　　244n6
　bedside and, 146, 161
　biological material and, 148, 159,
　　182–195, 199
　biotechnology and, 145–154, 160–
　　161, 165–169, 182, 186, 188, 192,
　　195
　central nervous system (CNS) research
　　and, 153, 169
　clinical trials and, 159, 170, 185, 189,
　　192, 194
　cognition and, 198–199
　collaboration and, 172, 175, 178, 181–
　　182, 192
　commercialization and, 10, 21, 46, 48,
　　61, 107, 146, 180–183, 244n6
　connection and, 192–195, 244n11
　depression and, 184–185
　diagnostics and, 146, 149, 159, 175,
　　178, 183–188, 191, 195

early-stage, 14, 49, 53, 58–59, 62f-63f, 67, 93, 105–106, 148, 157, 161, 184, 221, 230–231

entrepreneurs and, 147–148, 151, 166–173, 176–177, 181, 188–189, 247n1

environmental issues and, 123–124

epilepsy and, 153, 177, 195–196

epistemologies and, 145, 157–158, 180–181, 188, 190, 193, 198–199

ethics and, 195–201, 247n4

externalization and, 60–64

funding and, 56–60, 148, 188, 207

good science and, 51, 64–65, 75, 153

Humanitarian Device Exemption (HDE) and, 215–220

information engines and, 120

intellectual property (IP) and, 161, 166, 171–173, 196

investment and, 145, 149–158, 161, 176–178, 192, 195

late-stage, 53, 61, 63f, 109

licensing and, 147, 161, 166

Lynch and, 147–148, 167–178

market issues and, 145–148, 152–161, 165–166, 168, 171, 176, 179, 181, 183, 186, 188–189, 192, 196, 201, 247n1

mechanisms of action (MoA) and, 41, 70, 81, 154, 157–163, 188, 190

morals and, 227

neoliberalism and, 56, 181–182

neurotechnology and, 146–156, 162–163, 173–178, 181–184, 188–192, 195–201

Neurotechnology Investing and Partnering Conferences and, 146–149, 163–181, 183, 188, 195, 201

openness and, 59, 129, 133, 136–141, 246n12

partial subjects and, 195–201

partnerships and, 145, 168–170, 184

patents and, 154, 172, 175, 181, 196, 247n13

pharmaceuticals and, 146–162, 165–166, 168–170, 177–178, 181–187, 190, 192–197

pipelines and, 24, 36, 44, 46, 58, 60–61, 66, 81, 154–156, 169, 230, 238

policies for, 21–22

profit and, 146–147, 150, 152, 160, 168–169, 179, 196

proof of concept and, 59, 192

psychiatry and, 155–159, 174–175, 177, 182–185, 190–191, 200

research and development (R&D) and, 155–156, 169

risk and, 50–51, 149–161, 167–172, 182–183, 186, 190, 196–197, 201

socialities of translation and, 163–180

startups and, 146, 150–153, 157, 166–168, 173, 175, 177, 184, 188–189, 195–196

Stevenson-Wydler Technology Innovation Act and, 37

too much knowledge and, 160–162

translational neuroscience (TN) and, 145–151, 154, 157, 165, 171–173, 175, 178, 180–183, 186, 191–195, 200–201

US Food and Drug Administration (FDA) and, 160–161, 163

venture capital and, 146–148, 152–153, 157, 159, 166, 175, 181–183, 190, 247n3

working ambiguity and, 160

Insel, Thomas, 71–80, 244n9

Institute for Translational Medicine and Therapeutics, 43–44

Intellectual property (IP)

academic livelihoods and, 226–227

Bayh-Dole Act and, 22

copyright and, 61

financialization and, 58, 62f, 75, 77, 91–92

Intellectual property (IP) (cont.)
  innovation and, 161, 166, 171–173,
    196
  laboratories and, 107, 123, 140
  licensing and, 22, 36, 52–53, 58–63,
    66–69, 92, 102, 106, 140, 147, 161,
    166, 242n10, 244n6
  market issues and, 171–173
  "me-too drugs" and, 74–75
  mobility and, 173
  patents and, 22, 24, 36, 38, 51, 59, 61,
    65, 72–77, 81, 85, 92, 140, 154, 172,
    175, 181, 196, 234, 247n13
  political economy and, 36, 44, 46
  translational neuroscience (TN) and,
    22
  universities and, 107, 123, 140
Investment
  deal flow and, 117, 152–153
  early-stage, 14, 49, 53, 58–59, 62f-63f,
    67, 93, 105–106, 148, 157, 161, 184,
    221, 230–231
  entrepreneurs and, 1–2, 6–7, 13–14,
    52, 58–59, 119, 143, 145, 147, 151,
    166, 168, 176–177, 181, 188–189,
    232
  financialization and, 49–53, 57–59,
    62–64, 67–69, 80, 85–86, 93
  funding and, 7, 11, 33, 37, 43, 52, 62–
    63, 188
  growing expectations in, 57
  Humanitarian Device Exemption
    (HDE) and, 215–220
  initial public offerings (IPOs) and, 152
  innovation and, 145, 149–158, 161,
    176–178, 192, 195
  laboratories and, 98, 119
  lobbying and, 2, 6–7, 11
  mechanisms of action and, 157–163
  neurotechnology and, 1–2, 6–7, 10–
    14, 19, 23, 42, 51–52, 67, 86, 98,
    146–147, 150–155, 176–178, 181,
    188, 195, 201, 232

Neurotechnology Industry Report and, 6,
    23, 182
  Neurotechnology Investing and
    Partnering Conferences and, 146 (see
    also Neurotechnology Investing and
    Partnering Conferences)
  patients and, 204, 216, 220
  political economy and, 33–34, 36–37,
    42–43, 46–47
  research and development (R&D) and,
    26, 57, 59, 63, 69, 85, 93, 119, 156,
    232, 237
  risk and, 6, 10, 13–14, 26, 34, 50, 57,
    59, 62–63, 67, 69, 80, 85–86, 93,
    153–154, 161, 178, 216, 231, 237
  startups and, 6–7, 14, 50–52, 98, 146,
    150–153, 157, 166–168, 175, 188,
    195–196
  strategies for, 53, 59, 230–231
  translational neuroscience (TN) and,
    1, 4, 6–7, 10, 13–14, 19, 23, 26–27,
    241n2
  US Food and Drug Administration
    (FDA) and, 7, 12, 160–161, 214,
    216
  "valley of death" and, 11–12, 46
  venture capital and, 2, 5, 38, 46,
    50–51, 62f, 88, 99–102, 110, 117–
    118, 120, 142, 146–148, 152–157,
    159, 166, 175, 181–183, 190, 204,
    247n3
"Is Marketing the Enemy of
  Pharmaceutical Innovation?"
  (Applbaum), 88

Jansen Research & Development LLC,
    56
Jasanoff, Sheila, 203
Jawan, 210–214, 217, 220, 224
Jenkins, Janis, 187
Johns Hopkins, 52, 57–58
Johnson & Johnson, 52–54, 57–58
Jones, Dr., 79, 206–209, 226

*Journal of the American Medical Association*, 72
*Journal of Translational Medicine*, 204–205

Kahlon, Maninder, 109–114, 121
Kingery, W. David, 138
Knowledge constriction, 232–233
Knowledge problems, 84–89
Knowledge production, 17, 19, 21, 24, 94, 116, 145
Kraft, Alison, 48
Kuhn, Thomas, 235

Laboratories
  Alzheimer's and, 125–133
  anthropological studies and, 99, 110, 117, 119–120, 123, 126, 138, 146
  architecture and, 97, 101, 108–109, 121, 124–125, 130–142, 246nn11,12, 247n14
  "bench-to-bedside" approach and, 39–40, 43, 71, 84, 128, 131, 204–205, 207
  biotechnology and, 98–104, 110, 117, 123
  clinical trials and, 104, 142
  cognition and, 119–120, 245n8
  CoLaborator and, 102
  collaboration and, 102, 104–121, 125, 136–141
  commercialization and, 98, 107–108, 119–120, 123, 220, 245n8
  conferences as, 146
  entrepreneurs and, 99, 112–113, 116, 119, 122, 139–140, 143
  environmental issues and, 123–124, 130
  epistemologies and, 17–18, 97–99, 103, 107, 119, 130, 137–139
  ethics and, 98, 131, 142
  ethnographic analysis and, 16–19
  expertise mining and, 113–117, 120
  externalization and, 106–107, 111
  funding and, 41, 104–105, 116, 120, 125–126
  global health and, 16–19
  information engines and, 120
  intellectual property (IP) and, 107, 123, 140
  investment and, 98, 119
  licensing and, 102, 106, 140
  market issues and, 110, 114, 116, 119, 137, 140–141, 246n8
  Mission Bay and, 99–104, 108–109, 245n3
  morals and, 129, 138, 142–143
  Neurotechnology Investing and Partnering Conferences and, 98, 132, 146–148
  openness and, 59, 129, 133, 136–141, 246n12
  partnerships and, 3, 102–111, 118, 140
  pharmaceuticals and, 98, 102, 105–108, 118, 122, 128, 132, 143
  Post-Fordist models and, 115–116
  profit and, 106, 113
  proximity/intimacy and, 103–109
  research and development (R&D) and, 98, 102, 105, 107, 118–119
  risk and, 103, 105–107, 123
  science and technology studies (STS) and, 18, 146, 203
  translational neuroscience (TN) and, 98–99, 121–123, 128, 130, 132, 137–138, 141–142
  transparency and, 132–142, 247n14
  US National Institutes of Health (NIH) and, 103, 105–107, 114, 131
  vertical disintegration and, 93–96
  virtualization and, 55, 111–112
Latour, Bruno, 18–19, 128–130, 245n7
Lee, Peter, 247n13
Lefebvre, Henri, 141
Legibility of facts, 132
Leonelli, Sabina, 173

Leveraging
  commercialization and, 46, 89, 91, 118
  market issues and, 14, 26, 46, 49–50, 55–56, 60–63, 89, 91, 95, 118, 168, 186–187, 209, 230
  overleveraging and, 14, 89, 91, 95
  political economy and, 46
  translational neuroscience (TN) and, 14, 26
Libertarianism, 9, 61
Library of Congress, 40
Licensing
  Bayh-Dole Act and, 22, 67, 244n6
  innovation and, 147, 161, 166
  intellectual property (IP) and, 22, 36, 52–53, 58–63, 66–69, 92, 102, 106, 140, 147, 161, 166, 242n10, 244n6
  laboratories and, 102, 106, 140
  Office of Policy and, 242n10
  partnerships and, 52–56
  political economy and, 36, 242n10
Life Alert, 224–225
*Life Sciences Report*, 147–148
LinkedIn, 117–118, 168, 180
Linking, 192–193
Lock, Margaret, 41, 83, 194, 247n4
Lopez-Figuero, Manuel, 88–89
Lynch, Casey, 2, 5f, 6
Lynch, Zack. *See also* Neurotechnology Investing and Partnering Conferences
  financialization and, 96
  innovation and, 147–148, 167–178
  lobbying and, 2, 6–7, 11
  neurofuturism and, 4–6, 12
  NeuroInsights and, 2, 6–7, 23, 146–148
  *The Neuro Revolution* and, 3–4
  neurosociety and, 8–9
  Neurotechnology Industry Organization (NIO) and, 2, 7, 148
  political economy and, 30, 46

  as tracker, 5–6
  translational neuroscience (TN) and, 1–14, 30, 46, 96
Lyotard, Jean-François, 242n12
Lyrica, 68

Manipulability, 80, 194
Manuel, Luis, 190–191
Marginalization, 77, 182, 197, 204, 225, 233–235
Marincola, Francis, 204–205
Market issues
  bazaars and, 177–180, 188
  detethering and, 236–239
  disease and, 10, 51, 65, 88, 154, 188, 192, 196, 215, 217, 239
  economic bioethics and, 239
  financialization and, 49–54, 60–68, 71–72, 75–76, 86–88, 91–96, 245n15
  globalization and, 35–36, 54, 56, 98, 137, 140
  Humanitarian Device Exemption (HDE) and, 215–220
  innovation and, 145–148, 152–161, 165–166, 168, 171, 176, 179, 181, 183, 186, 188–189, 192, 196, 201, 247n1
  intellectual property (IP) and, 171–173
  knowledge problem and, 84–89
  laboratories and, 110, 114, 116, 119, 137, 140–141, 246n8
  leveraging and, 14, 26, 46, 49–50, 55–56, 60–63, 89, 91, 95, 118, 168, 186–187, 209, 230
  licensing and, 22, 36, 52–53, 58–63, 66–69, 92, 102, 106, 140, 147, 161, 166, 242n10, 244n6
  LinkedIn and, 117–118, 168, 180
  mental illness and, 12, 19, 54, 70, 72, 78, 81–82, 190
  "me-too drugs" and, 74–75
  opportunity and, 73–80

outsourcing and, 49, 56, 60–61, 63,
    66, 69, 94, 155, 231
partnerships and, 2–3, 7, 36–37, 42,
    49, 52, 62, 65, 67, 110, 140, 145–
    148, 166, 176, 192, 201, 238
patents and, 22, 24, 36, 38, 51, 59, 61,
    65, 72–77, 81, 85, 92, 140, 154, 172,
    175, 181, 196, 234, 247n13
patients and, 211, 214–217, 226–229,
    248n9
pharmaceuticals and, 3, 6–7, 27, 38,
    42, 46, 49–52, 60, 64–68, 71–76, 86–
    88, 93–95, 148, 156, 160–161, 166,
    168, 192, 196, 201, 216, 227, 229,
    232, 238, 248n9
political economy and, 32, 36–38, 41–
    42, 46
privatization and, 21–22, 25, 60–61,
    91–93, 95, 243n3, 245n15
productivity and, 19, 44, 69, 90, 99,
    160, 187–188, 243n16
profit and, 3, 10–12, 15, 23, 25 (see
    also Profit)
recalls and, 214, 248n6
role of connections and, 119
subjects/objects of neurotechnology
    and, 181–182
translational neuroscience (TN) and,
    230–232, 236
unmet needs and, 154, 188–189
vertical disintegration and, 93–96
"Markets, Myths, and Misrecognitions:
    Economic Populism in the
    Age of Financialization and
    Hyperinequality" (Ho), 32
Marshall, Chris, 160–162
Mauss, Marcel, 17
Mayo Clinic, 210–211
Mechanism of action (MoA)
    financialization and, 70, 81–82
    innovation and, 154, 157–163, 188,
        190
    political economy and, 41

Medicaid, 74
Medicare, 74
Medtronic, 211, 214–215, 249n11
Med-X Institute of Translation Research,
    35
Mental illness, 12, 19, 54, 70, 72, 78,
    81–82, 190
Merck, 54
"Me-too drugs," 74–75
Miku, 127–128, 130, 132–135, 137–139,
    141–142
Mirowski, Philip, 181
Mission Bay, 99–104, 108–109, 245n3
MIT, 85
Miu, John, 135–139
Mobility, 41, 121, 137, 173, 195
Mol, Annemarie, 200
Molecular medicine, 242n5
Molecular starting points, 220, 232–
    234
Morals
    destigmatization and, 244n8
    ethics and, 237 (see also Ethics)
    financialization and, 78, 188, 196,
        227
    Humanitarian Device Exemption
        (HDE) and, 215–220
    innovation and, 227
    laboratories and, 129, 138, 142–143
    libertarian, 9
    patients and, 216–217, 222
    research and development (R&D) and,
        129, 138, 142–143
    translational neuroscience (TN) and,
        4, 9–13, 30, 78, 138, 142–143, 229,
        237

National Center for Advancing
    Translational Sciences (NCATS),
    vii, 36–37, 56, 107, 120, 242nn7,10
National Institute of Mental Health
    (NIMH), 40, 71, 75, 77
Nature Reviews journal, 43–44

Neoliberalism
  financialization and, 22–25, 56, 61,
    66, 89, 92–93, 95
  innovation and, 56, 181–182
  research and development (R&D) and,
    21–25, 97–98, 140, 245nn15,16
  translational neuroscience (TN) and,
    22–25
  universities and, 21
Neurofuturism, 4–6, 12
Neurogenetics, 54
NeuroInsights, 2, 6–7, 23, 146–148
Neuroscience iMed, 55, 66, 111
Neurosociety, 8–9
Neurotechnology
  financialization and, 49–54, 59, 65,
    67, 75, 77, 85–89, 96
  innovation and, 146–156, 162–163,
    173–178, 181–184, 188–192, 195–201
  investment in, 1–2, 6–7, 10–14, 19,
    23, 42, 51–52, 67, 86, 98, 146–147,
    150–155, 176–178, 181, 188, 195,
    201, 232
  laboratories and, 98–99, 132
  patients and, 203, 208, 214
  political economy and, 42, 46
  risky investment in, 149–157
  translational neuroscience (TN) and,
    1–16, 21, 23, 26
Neurotechnology & Industry Partnering
    Conference, 191–192, 197
Neurotechnology Industry Organization
    (NIO), 2, 7, 146, 148
Neurotechnology Industry Report
    (NeuroInsights), 6, 23, 87, 182
Neurotechnology Investing and
    Partnering Conferences
  as bazaars, 177
  as closed events, 148
  financialization and, 2, 5, 15, 21, 42,
    54, 65, 88–89, 98, 132, 146–149,
    163, 173, 176–177, 181, 183, 188,
    195, 201, 203, 232, 234, 236

  innovation and, 146–149, 163–181,
    183, 188, 195, 201
  laboratories and, 98, 132, 146–148
  patients and, 203
  pharmaceuticals and, 146–148
  political economy and, 42
  research and development (R&D) and,
    146–148
  socialities of translation and, 163–
    180
  translational neuroscience (TN) and,
    2, 15, 21
  universities and, 146–148
Neuro: The New Brain Sciences and the
    Management of the Mind (Rose and
    Abi-Rached), 20
New England Journal of Medicine, 2–3,
    34
New Revolution, The: How Brain
    Science Is Changing Our World
    (Lynch), 3–4
Nguyen, Vinh-Kim, 41
Northwestern University, 68
Novartis, 54–55, 105–106
Nye, Jeffrey, 52–54, 57–58

Obesity, 82, 211
Office of Orphan Products Development
    (OOPD), 215
Openness
  architecture and, 102, 109, 123–125,
    132–141, 163–165, 246n11, 247n14
  importance of, 138–142
  modulating, 141
  research and development (R&D) and,
    59, 115–116, 129, 133, 136–141,
    246n12
  transparency and, 132–142, 247n14
Orphan Drug Designation, 215
Outsourcing, 49, 56, 60–61, 63, 66, 69,
    94, 155, 231
Overeating, 82
Overleveraging, 14, 89, 91, 95

Palin, Sarah, 30–34, 38, 241nn1,4
Parkinson's disease, 7, 211, 214
Partnerships
  Clinical and Translational Science
    Institute (CTSI) and, 109–111, 118
  CoLaborator and, 102
  financialization and, 49, 52–69,
    85, 90–91, 93, 230–231, 236, 238,
    242n10, 245n15
  funding and, 56–60
  innovation and, 145, 168–170, 184
  laboratories and, 3, 102–111, 118, 140
  licensing and, 52–56
  market issues and, 2–3, 7, 36–37, 42,
    49, 52, 62, 65, 67, 110, 140, 145–
    148, 166, 176, 192, 201, 238
  Neurotechnology Investing and
    Partnering Conferences and, 2, 5, 15,
    21, 42, 52, 54, 65, 98, 132, 146–149,
    163, 173, 176–177, 181, 183, 188,
    195, 201, 203, 232, 234, 236
  NIH hand-selecting of, 231
  pharmaceuticals and, 3, 7, 25, 49, 52–
    53, 59, 65–66, 90–91, 102, 105–106,
    109, 118, 168–169, 231, 238
  political economy and, 242n10
  psychiatry and, 52
  risk and, 14–15, 25, 49, 59, 62–63,
    66–67, 69, 94, 105, 107, 231, 236
  translational neuroscience (TN) and,
    3, 7, 15, 25–26
  universities and, 3, 102–111, 118,
    140
  vertical disintegration and, 93–96
  virtualization and, 55, 111–112
Pasteur, Louis, 129
Patents
  innovation and, 154, 172, 175, 181,
    196, 247n13
  intellectual property (IP) and, 22, 24,
    36, 38, 51, 59, 61, 65, 72–77, 81, 85,
    92, 140, 154, 172, 175, 181, 196,
    234, 247n13

"me-too drugs" and, 74–75
  political economy and, 36, 38
Patients
  academic livelihoods and, 226–227
  "bedside to bench" approach and,
    204–207
  "bench-to-bedside" approach and, 39–
    40, 43, 71, 84, 128, 131, 204–205,
    207
  biological material and, 218
  biotechnology and, 207, 213–220
  blood tests and, 184–186, 208
  Cleveland Clinic and, 14, 204, 210–
    213, 223, 226, 248n2
  clinical trials and, 215–216, 219–221
  collectives and, 43
  commercialization and, 214, 219
  crisis for, 42–47
  depression and, 211
  detethering and, 236–239
  diagnostics and, 208–209, 222
  discursive deconstruction and,
    207–208
  environmental effects and, 205–209,
    218
  epilepsy and, 209–214, 217–219, 223–
    225, 248n4
  epistemologies and, 205
  ethics and, 204, 213, 216, 219, 226–
    227, 229
  exemption and, 214–220
  Humanitarian Device Exemption
    (HDE) and, 215–220
  implants and, 7, 159, 186, 205, 211,
    213
  investment and, 204, 216, 220
  Life Alert and, 224–225
  marginalization of, x–xi, 197, 225,
    234–235
  market issues and, 211, 214–217, 226–
    229, 248n9
  Mayo Clinic and, 210–211
  morals and, 216–217, 222

Patients (cont.)
  narratives of, 210–214, 217, 220, 222–
    226, 224
  neurologist's narrative for, 206–209
  neurotechnology and, 203, 208, 214
  Neurotechnology Investing and
    Partnering Conferences and, 203
  personalized medicine and, 5, 84,
    185–186, 190
  pharmaceuticals and, 209, 216, 220,
    222, 227
  profit and, 209
  psychiatry and, 211, 222
  research and development (R&D) and,
    218, 221
  risk and, 212–222, 226–227
  safety and, 214–220, 222
  translational neuroscience (TN) and,
    203–206, 209, 212, 218–220, 227,
    229
  translational paradigm and, 203
  translational thinking and, 220–222
  translation issues and, 83–84
  US National Institutes of Health (NIH)
    and, 203–204, 217
  venture capital and, 204
"Patients-in-waiting," 40, 192, 209
Personalized medicine, 5, 84, 185–186,
  190
Pfizer
  Alzheimer's and, vii
  Centers for Therapeutic Innovation
    (CTI) and, 66, 104–108, 138,
    245n5
  China Basin and, 104, 109
  financialization and, 56, 60, 66, 68
  National Center for Advancing
    Translational Sciences (NCATS) and,
    vii
  University of California at San
    Francisco (UCSF) and, 104–108, 138,
    245n5
  Yale and, ix, vii

Pharmaceuticals, x, 241n2. See also
  Specific company
  Big Pharma and, 1, 50–52, 54, 75, 95,
    153–155, 182, 192, 209, 237
  biomarkers and, 41, 69, 71, 75, 80–82,
    191, 201, 234
  biopharmaceuticals and, 7, 15, 24,
    49–51, 56–61, 64, 67, 69, 79, 84–86,
    89, 91, 94, 105–106, 118, 143, 158–
    159, 169, 209, 230–232, 238, 243n5,
    248n9
  blockbuster drugs and, 68
  commercialization and, 2–3, 46, 65,
    89, 91
  epistemologies and, 237
  externalization and, 230–232
  financialization and, 49–61, 64–95,
    238
  funding and, 56–60
  innovation and, 146–162, 165–166,
    168–170, 177–178, 181–187, 190,
    192–197
  knowledge problems and, 84–89
  laboratories and, 98, 102, 105–108,
    118, 122, 128, 132, 143
  market issues and, 3, 6–7, 27, 38, 42,
    46, 49–52, 60, 64–68, 71–76, 86–88,
    93–95, 148, 156, 160–161, 166, 168,
    192, 196, 201, 216, 227, 229, 232,
    238, 248n9
  mental illness and, 12, 19, 54, 70, 72,
    78, 81–82, 190
  "me-too drugs" and, 74–75
  neuropharmacoepistemology and,
    157, 233–234
  Neurotechnology Investing and
    Partnering Conferences and,
    146–148
  partnerships and, 3, 7, 25, 49, 52–53,
    59, 65–66, 90–91, 102, 105–106, 109,
    118, 168–169, 231, 238
  patients and, 209, 216, 220, 222,
    227

pharmaceuticalization and, 27, 54,
    69, 73–74, 77, 82–83, 186, 209, 220,
    244n8
political economy and, 34, 38, 42,
    44–46
profit and, 3, 11, 25, 44, 50–51, 64–69,
    74, 81, 86, 88, 106, 150, 152, 168–
    169, 209, 231, 238
research and development (R&D)
    and, 15, 24, 49, 55–57, 69–70, 88–
    89, 93–94, 98, 102, 105, 156, 169,
    230–232
risk and, 13–14, 24–25, 34, 38, 50–51,
    57, 60, 64, 66–70, 75, 80, 83–95,
    105–107, 154–155, 159, 161, 167,
    170, 178, 182–183, 190, 196, 201,
    216, 218, 220, 222, 229–231
safety and, 11, 88, 102
science inside, 80–84
scientific uncertainty and, 73–80
translational neuroscience (TN) and,
    1–3, 6–7, 10, 13–15, 19, 22, 24–25,
    27
universities and, 2–3, 7, 13, 15, 22, 25,
    34, 49, 53, 55–56, 59, 64–69, 74, 89–
    94, 98, 102, 105–108, 143, 146, 151,
    161, 166, 181–182, 192, 195, 220,
    231–232, 243n5
US Food and Drug Administration
    (FDA) and, 7, 160–161, 214, 216,
    248nn6,9
US National Institutes of Health (NIH)
    and, 2–3, 35–36, 90, 105–107
vertical disintegration and, 93–96
Philanthropy, 43
Pipelines
detethering and, 238
financialization and, 52, 58, 60–61,
    66–67, 70–71, 80–81, 85
innovation and, 24, 36, 44, 46, 58, 60–
    61, 66, 81, 154–156, 169, 230, 238
political economy and, 33, 36, 43–44,
    46

research and development (R&D) and,
    24, 33, 36, 43–44, 46, 52, 58, 60–61,
    66–67, 70–71, 80–81, 85, 109, 231,
    235, 238
US National Institutes of Health (NIH)
    and, 33, 36, 43
Placebos, 71–72, 155, 158, 160, 219
Polanyi, Karl, 8
Political economy, x–xi
anthropology and, 32, 47
biotechnology and, 40–41
cognition and, 38, 41
collaboration and, 34–35, 38
commercialization and, 36, 44, 46, 48
diagnostics and, 36, 38–40, 46
entrepreneurs and, 34, 38
ethics and, 32
funding and, 29–37, 41, 43, 46–47
global economic crisis and, 33–34, 48
intellectual property (IP) and, 36, 44,
    46
investment and, 33–34, 36–37, 42–43,
    46–47
leveraging and, 46
licensing and, 36, 242n10
lobbying and, 2, 6–7, 11
Lynch and, 30, 46
market issues and, 32, 36–38, 41–42,
    46
Neurotechnology Investing and
    Partnering Conferences and, 42
partnerships and, 42, 242n10
patents and, 36, 38
pharmaceuticals and, 34, 38, 42,
    44–46
pipelines and, 33, 36, 43–44, 46
profit and, 44
research and development (R&D) and,
    30
risk and, 34, 38, 42, 241n2
semipermanent commercial
    architectures and, 30
taxpayers and, 30–33

Political economy (cont.)
  translational neuroscience (TN) and,
    29–30, 35, 38–42, 46–48
  translational science and medicine
    (TSM) and, 29–48
  US National Institutes of Health (NIH)
    and, 29, 32–37, 42–43, 242n7
  venture capital and, 38, 46
Populism, 31–33, 42
Post-Fordist models, 115–116
Powers, Joe, 3
Prescribing practices, 74–80
Presidio, 50, 151
Princeton University, 134
Privatization, 21–22, 25, 60–61, 91–93,
  95, 243n3, 245n15
Problematizability, 1, 7, 41, 159
Problem solving, 62f, 148, 190, 193
Productivity, 19, 44, 69, 90, 99, 160,
  187–188, 243n16
Profit
  CNS sector and, 86–87
  disease and, 10, 51, 65, 88, 154, 188,
    192, 196, 215, 217, 239
  financialization and, 51, 56, 64–69,
    74, 81, 86–88
  innovation and, 146–147, 150, 152,
    160, 168–169, 179, 196
  laboratories and, 106, 113
  leveraging and, 14, 26, 46, 49–50, 55–
    56, 60–63, 89, 91, 95, 118, 168, 186–
    187, 209, 230
  patients and, 209
  pharmaceuticals and, 3, 11, 25, 44,
    50–51, 64–69, 74, 81, 86, 88, 106,
    150, 152, 168–169, 209, 231, 238
  political economy and, 44
  psychiatric disorders and, 88–89, 182
  research and development (R&D) and,
    106, 113
  translational neuroscience (TN) and,
    3, 10–12, 15, 23, 25
Proof of concept, 59, 192

Psychiatry
  anthropology and, 79–80
  destigmatization and, 244n8
  financialization and, 52–55, 66, 70–
    81, 88, 182, 244nn8,9
  innovation and, 155–159, 174–175,
    177, 182–185, 190–191, 200
  neuropsychiatric models and, 14, 16,
    26, 156–157, 244n9
  next generation, 177
  partnerships and, 52
  patients and, 211, 222
  profit and, 88–89, 182
  Rorschach and, 174
  Stein and, 222
  translational neuroscience (TN) and,
    2–3, 6–16, 19, 21–27, 39
Psychology, 4, 54, 174
Psychopharmaceuticals
  financialization and, 51, 54, 65, 69–
    72, 76–89, 91, 93, 95
  innovation and, 154–155, 180–190,
    201
  laboratories and, 128
  retreat from, 69–72
  science inside, 80–84
  translational neuroscience (TN) and,
    2, 19
"Pychodiagnostic Plates" (Rorschach),
  174

Rabinow, Paul, 85, 123
Rational choice theory, 176–177
Recalls, 214, 248n6
Requests for proposals (RFPs), 41–42
Research and development (R&D)
  abandonment in, 53–55, 57, 69–72,
    86, 156, 243n2
  Alzheimer's and, 125–133
  ambiguity and, 72, 75–76, 89, 159–
    161, 213, 244n7
  architecture and, 97, 101, 108–109,
    121, 124–125, 130–142, 247n14

biomarkers and, 41, 69, 71, 75, 80–82, 191, 201, 234

central nervous system (CNS) and, 7, 10–13, 23, 26, 47, 53–57, 64–65, 69–71, 76, 84–89, 95, 153, 169, 241n2

Clinical and Translational Science Institute (CTSI) and, 109–122, 138, 142, 220, 245n5

clinical trials and, 104, 142 (*see also* Clinical trials)

collaboration and, 102, 104–115 (*see also* Collaboration)

contoured actions and, 121–122

cost of, 57, 105–106

"decade of the brain" and, 13, 40

downsizing and, 54, 56–57, 69, 231

early-stage, 14, 49, 53, 58–59, 62f-63f, 67, 93, 105–106, 148, 157, 161, 184, 221, 230–231

entrepreneurs and, 59, 169, 232

environmental issues and, 123–124

expertise mining and, 113–117, 120

externalization of, 55, 59, 61–64, 69, 93–94, 230–232

financialization and, 22–25, 49, 55–64, 69–70, 85, 88–89, 93–96, 238, 243n2

focus on good science and, 51, 64–65, 75, 153

funding and, 7, 11, 25, 29–37, 41, 43, 56–63, 78, 92, 104–105, 116, 120, 125–126, 207, 217, 226, 232, 244n9, 245n15

information engines and, 120

innovation and, 155–156, 169

internal, 55–57, 69, 94, 105, 243n2

investment in, 26, 57, 59, 63, 69, 85, 93, 119, 156, 232, 237

joint agreements and, 63

knowledge problems and, 84–89

knowledge production and, 17, 19, 21, 24, 94, 116, 145

laboratories and, 98, 102, 105, 107, 118–119

leveraging and, 14, 26, 46, 49–50, 55–56, 60–63, 89, 91, 95, 118, 168, 186–187, 209, 230

lowering costs of, 57

market issues and, 110, 114, 116, 119, 137, 140–141

mechanisms of action and, 70, 81–82, 157–163

Mission Bay and, 99–104, 108–109, 245n3

morals and, 129, 138, 142–143

neoliberalism and, 21–25, 97–98, 140, 245nn15,16

Neurotechnology Investing and Partnering Conferences and, 146–148

openness and, 59, 115–116, 129, 133, 136–141, 246n12

opportunity and, 73–80

outsourcing, 49, 56, 63, 94, 155

patients and, 218, 221

pharmaceuticals and, 15, 24, 49, 55–57, 69–70, 88–89, 93–94, 98, 102, 105, 156, 169, 230–232, 241n2

pipeline of, 24, 33, 36, 43–44, 46, 52, 58, 60–61, 66–67, 70–71, 80–81, 85, 109, 231, 235,

political economy and, 30

Post-Fordist models and, 115–116

problem solving and, 62f, 148, 190, 193

profit and, 106, 113

risk and, 11, 15–16, 23, 25–26, 34, 49, 57, 64, 66–67, 69, 71, 80, 85–86, 91–95, 105–107, 123, 154, 182–183, 186, 212, 217–218, 222, 226, 230, 236

role of connections and, 119

science and technology studies (STS) and, 18, 146, 203

scientific uncertainty and, 73–80

Research and development (R&D)
(cont.)
startups and, 98–99, 102, 109–110
Systems of National Accounts (SNA)
and, 118–119
translational neuroscience (TN) and,
10, 15–16, 24, 26, 237
transparency and, 132–142, 247n14
United Nations and, 118–119
universities and, 2–3, 7, 11–15, 21–
22, 25, 29–30, 33–41, 49–50, 55–58,
62–69, 74, 90–99, 103–108, 112–
113, 118–123, 126, 131, 134–135,
142, 146, 151, 166, 171, 173, 181–
182, 204, 226, 231–233, 242nn5,12,
245nn8,15
US National Institutes of Health (NIH)
and, 2–3, 29, 34–36, 43, 50, 55–56,
64, 105–106, 131, 203–204, 217
venture capital and, 99–102, 110, 117–
118, 120, 142
vertical disintegration and, 93–96
virtualization and, 55, 111–112
Research cores, 102, 113–115
Research Domain Criteria (RDoC), 83
Risk
constructivism and, 41, 193, 217–219
derisking mechanisms and, 14, 67,
69, 95
diagnostics and, 183–184
exemption and, 214–220
externalization and, 60–64, 230–232,
237
financialization and, 49–53, 57, 59–
71, 75, 80, 83–95
innovation and, 50–51, 149–161, 167–
172, 182–183, 186, 190, 196–197,
201
investment and, 6, 10, 13–14, 26, 34,
50, 57, 59, 62–63, 67, 69, 80, 85–86,
93, 153–154, 161, 178, 216, 231, 237
knowledge problems and, 84–89
laboratories and, 103, 105–107, 123

opportunity and, 73–80
outsourcing and, 49, 56, 60–61, 63,
66, 69, 94, 155, 231
partnerships and, 14–15, 25, 49, 59,
62–63, 66–67, 69, 94, 105, 107, 231,
236
patients and, 212–222, 226–227
pharmaceuticals and, 13–14, 24–25,
34, 38, 50–51, 57, 60, 64, 66–70, 75,
80, 83–95, 105–107, 154–155, 159,
161, 167, 170, 178, 182–183, 190,
196, 201, 216–222, 229–231
political economy and, 34, 38, 42,
241n2
research and development (R&D) and,
11, 15–16, 23, 25–26, 34, 49, 57,
64, 66–67, 69, 71, 80, 85–86, 91–95,
105–107, 123, 154, 182–183, 186,
212, 217–218, 222, 226, 230, 236
scientific uncertainty and, 73–80
scientific vs. financial, 84–89
shareholder, 13, 16, 49, 57, 59, 64, 66,
93, 229–231
translational neuroscience (TN) and,
4, 6, 10–16, 23–26, 248n10
translational thinking and, 220–222
vertical disintegration and, 93–96
Risk and Culture (Douglas and
Wildavsky), 217
Roberts, Linda, 188–190, 198
Rorschach, Hermann, 174, 198
Rose, Nikolas, 20, 194

Safety
clinical trials and, 10–12, 58, 61, 71,
81, 85–86, 88, 104, 142, 159, 170,
185, 189, 192, 194, 215–221
DBS devices and, 214, 217, 219
HDE approval and, 215
patients and, 214–220, 222
pharmaceuticals and, 11, 88, 102
translational neuroscience (TN) and,
11, 22

Sandel, Michael, 227
San Francisco Bay Area, 6, 50, 99, 101
Sanofi-Aventis, 55–56
Sara, 126, 135
Schiff, Steven, 213
Schizophrenia, 54, 79, 154, 174, 195, 198–199
Scholarships, 61–63
Schumpter, Joseph, 8
Science and technology studies (STS), 18, 146, 203
Screening, 82, 184–185, 197
*Second Life* (Boellstorff), 117
Semipermanent commercial architectures, 30
Shana, 210–214, 217, 220, 224
Shanghai Jiatong University, 35
Shapin, Steven, 181–182, 246n9
Shareholders
  growing expectations of, 57
  risk and, 13, 16, 49, 57, 59, 64, 66, 93, 229–231
Sharp, Leslie, 205, 247n4
Silicon Valley, 2, 6, 46, 50, 78, 99, 136, 150–151, 171, 180, 214
Silverman, Richard, 68
Simmel, Georg, 141, 247n14
Simon, Dr., 155–159
SkyBridge Diagnostics, 184
Society for Neuroscience Conference, 73, 150
Sociotechnical ensemble, 8
Solomon, Miriam, 47, 203
Soteriology, 4, 208
Specificity, 22–23, 46, 95, 127, 149, 160
Stanford, 1, 21, 91, 110, 134, 139, 151
Startups
  entrepreneurs and, 6, 52, 92, 99, 102, 110, 151, 166, 168, 173, 188–189
  financialization and, 50–52, 62f, 85, 92
  innovation and, 146, 150–153, 157, 166–168, 173, 175, 177, 184, 188–189, 195–196

  investment and, 6–7, 14, 50–52, 98, 146, 150–153, 157, 166–168, 175, 188, 195–196
  research and development (R&D) and, 98–99, 102, 109–110
Stein, Dr., 78–80, 222
Stevenson-Wydler Technology Innovation Act, 37
Sunder Rajan, Kaushik, 60, 92, 173, 244n12
Systems of National Accounts (SNA), 118–119

Taxpayers, 30–33
Technoeconomic wave, 8–9
Technology Accelerator Fund, 57–58
Technophenomenon, 82
Tethering, 1, 9, 78, 145, 183, 193, 199, 201, 236–239
Thiel, Herry, 192–193, 197
Tiffany, 127, 130, 135
Trackers, 5–6
Transitivity, 193–194
"Translational and Clinical Science: Time for a New Vision" (Zerhouni), 2–3, 34
Translational neuroscience (TN)
  Alzheimer's and, 10–11, 231
  anthropology and, 6, 13, 16–19, 24, 26, 96, 151
  biotechnology and, 1–7, 14–15, 19, 22–23, 26, 241n2
  central nervous system (CNS) and, 7, 10–13, 23, 26, 241n2
  clinical side of, 204–206
  clinical trials and, 10–12
  cognition and, 17, 19, 26
  collaboration and, 17, 19, 236
  commensurability and, 235–236
  commercialization and, 2–3, 7, 9–11, 21–25, 91, 245n8
  concept of, 2–3, 38–41
  depression and, 10, 231

Translational neuroscience (TN) (cont.)
diagnostics and, 2, 5, 12, 16, 27
entrepreneurs and, 1–2, 6–7, 13–14, 22–23
environmental issues and, 123–124
epilepsy and, 10
epistemologies and, 9, 15, 27, 232, 236
ethics and, 1–2, 14–16, 19–20, 26
ethnographic analysis of, 3, 7, 9–16, 19–26
externalization and, 15, 25
financialization and, 14, 22–26, 49, 51, 55–57, 59, 66–70, 74, 77–78, 81–82, 85, 91, 93–96, 230–232
funding and, 7, 11, 23, 25, 231–232
global health and, 16–19
globalization and, 35–36
goals in mind and, 127–130
innovation and, 145–151, 154, 157, 165, 171–173, 175, 178, 180–183, 186, 191–195, 200–201
insights into, 14–16
intellectual property (IP) and, 22
investment and, 1, 4, 6–7, 10, 13–14, 19, 23, 26–27, 241n2
laboratories and, 98–99, 121–123, 128, 130, 132, 137–138, 141–142
leveraging and, 14, 26
Lynch and, 1–14, 30, 46, 96
market issues and, 2–3, 6–12, 16, 21–27, 230–232, 236
mechanisms of action and, 41, 70, 81, 154, 157–163, 188, 190
as mining for gold, 42
molecular starting points of, 220, 232–234
morals and, 4, 9–13, 30, 78, 138, 142–143, 229, 237
neoliberalism and, 22–25
NeuroInsights and, 2, 6–7, 23, 146–148
neuropharmacoepistemology and, 233–234
neurotechnology and, 1–16, 21, 23, 26
Neurotechnology Investing and Partnering Conferences and, 2, 5, 15, 21
notion of discovery and, 39–40
partnerships and, 3, 7, 15, 25–26
patients and, 40, 192, 203–206, 209, 212, 218–220, 227
pharmaceuticals and, 1–3, 6–7, 10, 13–15, 19, 22, 24–25, 27
political economy and, 29–30, 35, 38–42, 46–48
presumptions of translatability and, 47
profit and, 3, 10–12, 15, 23, 25
psychiatry and, 7, 10, 13–14, 16, 19, 26, 39
research and development (R&D) and, 10, 15–16, 24, 26, 237
risk and, 4, 6, 10–16, 23–26, 248n10
safety and, 11, 22
scientific excess and, 41–42
sociality of, 181–182
as solution from failures, 42–47
spaces of investigation and, 13–14
US National Institutes of Health (NIH) and, 2–3
venture capital and, 2, 5
Zerhouni on, 2–3
Translational oncology, 16, 33–34, 152, 213, 232
"Translational Partnering and Investing Opportunities" panel, 175
Translational science and medicine (TSM)
"bench-to-bedside" approach and, 39–40, 43, 71, 84, 128, 131, 204–205, 207
biomarkers and, 41, 69, 71, 75, 80–82, 191, 201, 234
cost of, 230
detethering and, 236–239
diagnostics and, 36, 38–40, 46

financialization and, 49, 230–232 (*see also* Financialization)
history of, 33–38
laboratories and, 97–99 (*see also* Laboratories)
as long game, 230
many definitions of, vii-viii
mechanisms of action and, 70, 81–82, 157–163
molecular starting points of, 220, 232–234
multivalency of, 203
National Center for Advancing Translational Sciences (NCATS) and, vii, 36–37, 56, 107, 120, 242nn7,10
novelty of, 47–48
political economy and, x-xi, 29–48
role of connections and, 119
semipermanent commercial architectures and, 30
as solution from failures, 42–47
success of, 229–230
taxpayer as shareholder and, 30–33
vertical disintegration and, 93–96
Translational thinking, 39, 58
environment for, 142–143
morals and, 142–143
neuropharmacoepistemology and, 233–234
patients and, 220–222
research and development (R&D) and, 118, 128, 138–139, 142–143
risk and, 220–222
Translation issues, 83–84
Transparency
architecture and, 102, 109, 123–125, 130–142, 163–165, 246n11, 247n14
universities and, 132–142, 247n14
Traweek, Sharon, 246n9
Trinity University, 139
Tufts University, 35
Tyfield, David, 23

United Nations Statistical Commission, 118–119
Universities. *See also* Specific university
architectural effects and, 124–125, 130–142, 247n14
Bayh-Dole Act and, 67
collaboration and, 102, 104–121, 125, 136–141
commercialization and, 2–3, 11, 21–22, 50, 65, 67, 89–92, 98, 120, 123, 181, 245n8
entrepreneurs and, 89–93
environmental issues and, 123–124
expertise mining and, 113–117, 120
externalization and, 60–64, 106–107, 111
focus on good science and, 51, 64–65, 75, 153
funding and, 7, 11, 29, 34–35, 37, 41, 47, 65, 92, 104–105, 116, 120, 125–126, 245n15
information engines and, 120
intellectual property (IP) and, 107, 123, 140
listening to architectures and, 124–125
neoliberalism and, 21
Neurotechnology Investing and Partnering Conferences and, 146–148
openness and, 59, 129, 133, 136–141, 246n12
partnerships and, 3, 102–111, 118, 140
Pfizer and, vii, 104, 107–108, 138, 245n5
pharmaceuticals and, 2–3, 7, 13, 15, 22, 25, 34, 49, 53, 55–56, 59, 64–69, 74, 89–94, 98, 102, 105–108, 143, 146, 151, 161, 166, 181–182, 192, 195, 231–232, 243n5
Post-Fordist models and, 115–116
proximity/intimacy and, 103–109

Universities (cont.)
  research and development (R&D) and,
    2–3, 7, 11–15, 21–22, 25, 29–30, 33–
    41, 49–50, 55–58, 62–69, 74, 90–99,
    103–108, 112–113, 118–123, 126,
    131, 134–135, 142, 146, 151, 166,
    171, 173, 181–182, 204, 226, 231–
    233, 242nn5,12, 245nn8,15
  risk and, 60–69
  transparency and, 132–142, 247n14
  US National Institutes of Health (NIH)
    and, 2–3, 34–35, 37, 50, 55–56, 64,
    66, 103–107, 131
  vertical disintegration and, 93–96
  virtualization and, 55, 111–112
University of Alabama, 35
University of California at Berkeley, 13,
  85–86
University of California at Los Angeles
  (UCLA), 35
University of California at San Diego
  (UCSD), 39
University of California at San Francisco
  (UCSF)
  Bayer and, 102
  Clinical and Translational Science
    Institute (CTSI) and, 109–122, 138,
    142, 220, 245n5
  collaboration and, 109–121
  expertise mining and, 113–117, 120
  free market and, 110
  Genentech and, 101, 167
  Mission Bay and, 98–104, 108–109,
    245n3
  openness and, 140
  partnerships and, 110
  Pfizer and, 104–108, 138, 245n5
  pharmaceuticals and, 98, 102, 108, 220
  Profiles and, 113–115, 117–118, 122,
    245n6
  research cores of, 102, 113–114
  Weill Institute for Neurosciences and,
    101

University of Nebraska, 39, 41
University of Pennsylvania, 3, 44
Unmet needs, 154, 188–189
US Food and Drug Administration
  (FDA)
  clinical trials and, 12, 160
  Deep Brain Stimulation (DBS) and, 214
  Humanitarian Device Exemption
    (HDE) and, 215–220
  innovation and, 160–161, 163
  Medtronic recall and, 214, 248n6
  Neurotechnology Industry
    Organization (NIO) and, 7
  Orphan Drug Designation and, 215
  pharmaceuticals and, 7, 160–161, 214,
    216, 248nn6,9
US National Institutes of Health (NIH),
  viii
  Alzheimer's disease and, vii
  awards program of, ix
  commercialization and, vii, 2–3, 36,
    50, 105
  Discovering New Therapeutic Uses for
    Existing Molecules and, vii, 56, 105,
    107
  financialization and, 50, 56, 64, 66, 90
  funding and, 34–36, 105, 217
  Insel and, 71–80, 244n9
  laboratories and, 103, 105–107, 114,
    131
  Marincola and, 204–205
  National Center for Advancing
    Translational Sciences (NCATS) and,
    vii, 36, 56, 106–107, 242n7
  NIH Roadmap and, 34, 42, 90
  patients and, 203–204, 217
  pharmaceuticals and, 2–3, 35–36, 90,
    105–107
  political economy and, 29, 32–37, 42–
    43, 242n7
  research and development (R&D) and,
    2–3, 29, 34–36, 43, 50, 55–56, 64,
    105–106, 131, 203–204, 217

Roadmap for Medical Research and, 34
translational pipeline and, 33, 36, 43
universities and, 2–3, 34–35, 37, 50,
    55–56, 64, 66, 103–107, 131
Zerhouni and, 2–3, 29, 34, 55, 204

"Valley of death," 11–12, 46
Veblen, Thorstein, 21
Venture capital
  deal flow and, 117, 152–153
  financialization and, 50–51, 62f, 88
  innovation and, 146–148, 152–153,
      157, 159, 166, 175, 181–183, 190,
      247n3
  mechanisms of action and, 157–163
  patients and, 204
  political economy and, 38, 46
  research and development (R&D) and,
      99–102, 110, 117–118, 120, 142
  translational neuroscience (TN) and,
      2, 5
Vertical disintegration, 93–96
Virtualization, 55, 111–112

Washington University, 85
Webster, Andrew, 140
Weill Institute for Neurosciences, 101
Wellcome Trust Scottish Translational
    Medicine and Therapeutics
    Initiative, 35
White, Suzanne, 171–173, 180
Wildavsky, Aaron, 217
Wynne, Bryan, 245n8

Yale, vii, ix, 155
Yocca, Frank D., 55

Zerhouni, Elias, 2–3, 29, 34, 55, 204

Printed in the United States
by Baker & Taylor Publisher Services